2 50

FLOWERS OF THE SOUTHWEST MESAS

By Pauline M. Patraw

Illustrated by Jeanne R. Janish

Editor: Earl Jackson

SOUTHWEST PARKS AND MONUMENTS ASSOCIATION

339 South Broad Street

Box 1562, Globe, Arizona 85501

TABLE OF CONTENTS

Map of area with which this book is concerned 4
Preface 5
The Pinyon-Juniper Woodland 6, 7
Tree and Tree-like Shrubs 8-16
White Flowers 17-36
Pinkish and Reddish Flowers 37-55
Yellow Flowers 56-86
Greenish Flowers 87-91
Bluish and Purplish Flowers 92-107
Literature consulted 108
Index 109

PLANT FAMILIES REPRESENTED

TREES and TREE-LIKE SHRUBS

PINE FAMILY (*Pinaceae*) 8

CYPRESS FAMILY (*Cupressaceae*) 9, 10

WILLOW FAMILY (*Salicaceae*) 12

BEECH FAMILY (*Fagaceae*) 13

MAPLE FAMILY (*Aceraceae*) 14, 15

OLIVE FAMILY (*Oleaceae*) 16

SHRUBS HERBS

JOINTFIR FAMILY (*Ephedraceae*) 87

SPIDERWORT FAMILY (*Commelinaceae*) 92

LILY FAMILY (*Lilaceae*) 17, 18, 87

AMARYLLIS FAMILY (*Amaryllidaceae*) 56

ORCHID FAMILY (*Orchidaceae*) 88

BUCKWHEAT FAMILY (*Polygonaceae*) 38, 57

GOOSEFOOT FAMILY (*Chenopodiaceae*) 58, 89

FOUR-O'CLOCK FAMILY (*Nyctaginaceae*) 39, 93

PORTULACA FAMILY (*Portulacaceae*) ..40

BUTTERCUP FAMILY (*Ranunculaceae*) 19, 40, 59, 94

BARBERRY FAMILY (*Berberidaceae*) ..60

POPPY FAMILY (*Papaveraceae*) 20, 61

MUSTARD FAMILY (*Cruciferae*) 41, 62, 63

CAPER FAMILY (*Capparidaceae*) 59, 95

SAXIFRAGE FAMILY (*Saxifragaceae*) 21, 22, 64

ROSE FAMILY (*Rosaceae*) 23-27, 42, 65

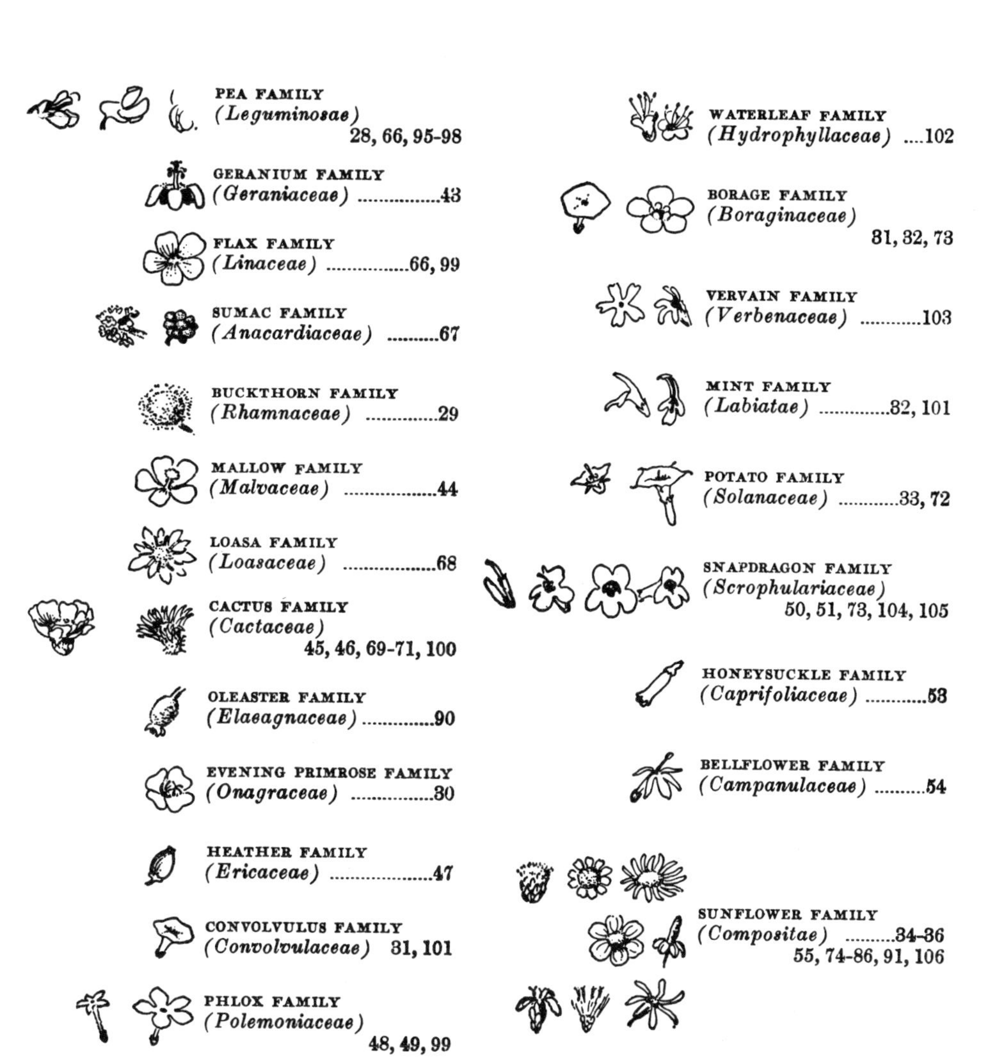

PEA FAMILY
(Leguminosae) 28, 66, 95-98

GERANIUM FAMILY
(Geraniaceae) 43

FLAX FAMILY
(Linaceae) 66, 99

SUMAC FAMILY
(Anacardiaceae) 67

BUCKTHORN FAMILY
(Rhamnaceae) 29

MALLOW FAMILY
(Malvaceae) 44

LOASA FAMILY
(Loasaceae) 68

CACTUS FAMILY
(Cactaceae) 45, 46, 69-71, 100

OLEASTER FAMILY
(Elaeagnaceae) 90

EVENING PRIMROSE FAMILY
(Onagraceae) 30

HEATHER FAMILY
(Ericaceae) 47

CONVOLVULUS FAMILY
(Convolvulaceae) 31, 101

PHLOX FAMILY
(Polemoniaceae) 48, 49, 99

WATERLEAF FAMILY
(Hydrophyllaceae) 102

BORAGE FAMILY
(Boraginaceae) 31, 32, 73

VERVAIN FAMILY
(Verbenaceae) 103

MINT FAMILY
(Labiatae) 32, 101

POTATO FAMILY
(Solanaceae) 33, 72

SNAPDRAGON FAMILY
(Scrophulariaceae) 50, 51, 73, 104, 105

HONEYSUCKLE FAMILY
(Caprifoliaceae) 53

BELLFLOWER FAMILY
(Campanulaceae) 54

SUNFLOWER FAMILY
(Compositae) 34-36 55, 74-86, 91, 106

DON'T WORRY ABOUT THE VERY FEW TECHNICAL TERMS USED —they are all explained in the following drawing showing flower parts.

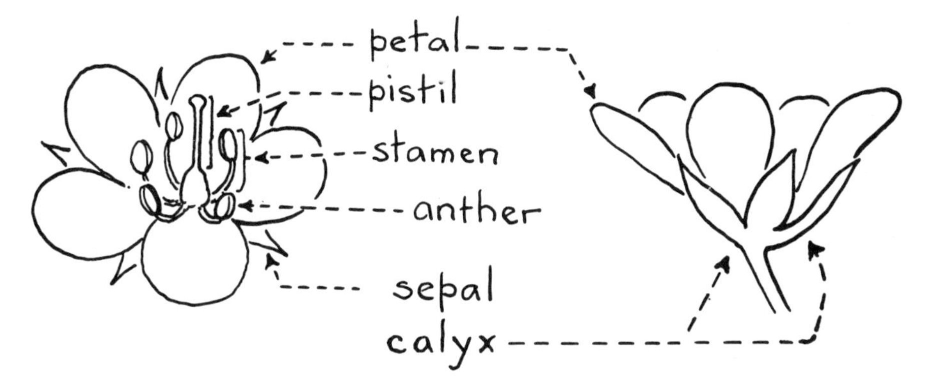

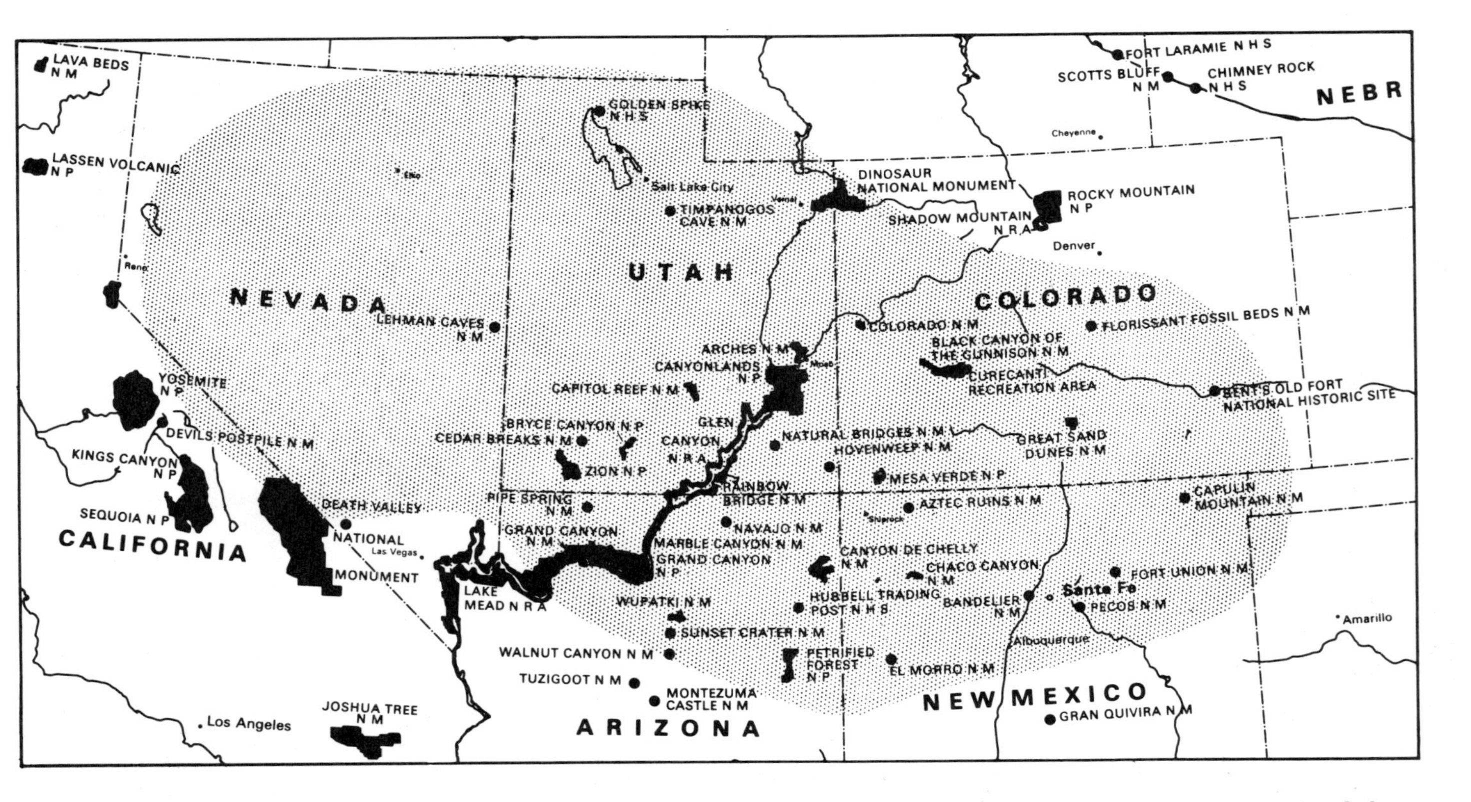

Flowers described in this book are found in the general area covered on the map with shading. For flowers growing below 3,000 feet elevation see the companion book to this one, "Flowers of the Southwest Deserts," by Dodge. Mountain flowers above 7,000 feet are contained in "Flowers of the Southwest Mountains," by Arnberger.

PREFACE

The purpose of this booklet is to make available to the interested layman names of, and information about, some of the common and conspicuous plants of the Upper Sonoran Zone, the "Pinyon-Juniper Woodland" of the Southwest.

The plants here described, with the exception of the trees and a few tree-like shrubs, are arranged according to flower color. Within these groups, as far as possible, the arrangement is according to the standard, systematic order of families as given in "Arizona Flora" by Kearney and Peebles.

To facilitate identification, small drawings of flower types of each family are given in the Table of Contents.

The flower color arrangement is as follows:

(1) White—includes white, whitish sometimes tinged with other colors, and cream.

(2) Pinkish and reddish—includes all shades of pink, rose, coral, and red.

(3) Yellow—all shades of yellow, sometimes marked or tinged with other colors.

(4) Greenish—flowers predominantly green.

(5) Bluish and purplish—includes all shades of blue, orchid, violet, purple, and magenta.

The color arrangement is not logical from the botanical standpoint, since color is seldom an identifying characteristic; but, due to the popular nature of this book, it is believed that this method will be most usable for the amateur botanist. Those wishing technical botanical keys and descriptions of plants of this area are referred to "Arizona Flora" by Kearney and Peebles (from which volume much information has been taken), "Flora of New Mexico" by Wooton and Standley, and "Flora of Utah and Nevada" by Tidestrom.

It is regrettable that space limitation does not permit descriptions of all of the most common plants of the Pinyon-Juniper Belt, but an attempt has been made to include the most outstanding plants in which the flower lover may be interested.

Every effort has been made to insure accuracy of the scientific names included in this book. Nearly all plants have been identified by the United States National Herbarium. The cacti have been identified by Dr. Lyman Benson while the book has been checked for botanical accuracy by Mr. Robert H. Peebles, who has been most generous in his help. The author is also indebted to Dr. Thomas H. Kearney for his criticism of the plant list chosen for this book; to Dr. Harold C. Bryant, Superintendent of Grand Canyon National Park; and Mr. Louis Schellbach, III, Park Naturalist of Grand Canyon National Park, for use of herbarium specimens; and to the superintendents of all the National Parks and Monuments of the Southwest for the plant lists of their areas.

It has been a delightful and stimulating experience to work with Jeanne Russell Janish who is well informed in botany as well as being an accurate and talented illustrator.

PAULINE MEAD PATRAW
Santa Fe, New Mexico

September 11, 1952

THE PINYON-JUNIPER WOODLAND

Across great stretches of country in the southwestern United States, between the desert of giant cactus and the mountains clothed with pine, spruce and fir, lies the land of the pinyon, juniper and sagebrush. Here are foothills of the mountain ranges, open plains cut with arroyos, mesas blue in the distances, and rocky canyons.

This is a land with a flavor all its own: a land of brilliant sunshine and cool breezes, where the fragrance of sagebrush is strong after a rain and where the sweet odor of pinyon smoke hangs in the air. Rainfall is light in this country. The trees are small and scrubby as though stunted, and generally grow some distance apart, making visible wide expanses of land and sky.

RANGE OF THE ZONE

The Sagebrush-Pinyon-Juniper Belt, known as the Upper Sonoran Zone, extends from Colorado and Utah south to central Arizona and New Mexico. It reaches to some extent on the west into eastern Nevada and California, and on the east into western Kansas, Oklahoma and Texas. (See map on page 4.) The altitudinal range of this zone is from about 4,500 to 7,500 feet and its limits vary with differences in exposure and moisture conditions.

PLANT ASSOCIATIONS

The lower elevations of the Upper Sonoran Zone are dominated by big sagebrush *(Artemisia tridentata)*, an indicator of good deep, well-drained soil, free from alkali. This typically southwestern shrub grows from 2 to 7 feet high, varying with soil conditions. It often covers large areas in pure stands. The early pioneers could easily spot soil that would make good farmland by the type of sagebrush growth.

The pretty yellow-flowered rabbitbrush or "chamisa" *(Chrysothamnus)* is more resistant to drought than is sagebrush, and prefers lighter, alkali-free soil. It is common all through the Pinyon-Juniper Belt.

Where rainfall is less, the soil more impervious, and salt content higher, the fourwing saltbush or "shadscale" *(Atriplex canescens)* grows. "Black" greasewood *(Sarcobatus vermiculatus)* dominates areas where soil is alkaline. Greasewood and saltbush are often found together.

At slightly higher elevations above the true sagebrush stands, junipers appear in scattered groups. Higher up they occur with low growing pinyon. Still higher, where rainfall is heavier, the pinyons form large, almost pure stands and are considerably taller. Such forests of pinyon occur in Mesa Verde National Park, in the Grand Canyon region, and around Flagstaff, Arizona.

In the upper reaches of the Pinyon-Juniper Belt, the Rocky Mountain juniper *(Juniperus scopulorum)*, mountain-mahogany *(Cercocarpus)* and the Gambel oak *(Quercus gambelii)* appear.

All through the Pinyon-Juniper Belt are many species of flowering shrubs and herbs in remarkably large numbers, considering the scarcity of rainfall and the short growing season.

PLANT ADAPTATIONS

In this semi-arid region, often referred to as the "semi-desert," plants have adapted themselves in several ways. Leaf surface is often reduced to prevent excessive evaporation, as in the junipers which have scale-like leaves, and the cacti and ephedras which have only rudimentary leaves or none at all. Many semi-desert plants develop extensive root systems to take up all available moisture; i.e., the bush morning-glory *(Ipomoea leptophylla)* and the sagebrush. Some plants such as cacti, store water in their tissues; and others have thick "skins," sometimes covered with wax, to prevent evaporation, as the leaves of the yuccas, and the pinyon needles. Most semi-desert plants are dull green or gray-green, often silvery, because of protective covering of hair or wool: for example, the sagebrush and the golden crownbeard *(Verbesina encelioides)*.

NATIONAL PARKS AND MONUMENTS AS WILD FLOWER SANCTUARIES

Throughout our country the government, through the National Park Service, protects areas that are outstanding because of their scenic, historic, prehistoric or scientific features. These areas are the National Parks and Monuments (see map), and are frequently referred to in this book when discussing distribution of the plants.

These, of course, are not the only localities where the plants occur, but in these areas plants and animals live undisturbed, under natural conditions. These are places where tourists are most apt to find wild flowers in abundance; and here, too, plants are studied and listed. Anyone wishing to learn more about plants when visiting the National Parks and Monuments should ask the rangers or naturalists. Many National Park areas have herbarium specimens for reference and study, and nearly all have plant lists. Some of the National Monuments have quite complete plant lists of the surrounding country. This is true of Aztec Ruins, Chaco Canyon, and Wupatki National Monuments. When it is stated that a plant grows in one of these Monuments it also may be considered to occur in the vicinity.

There are vast stretches of wilderness in the pinyon-juniper country. This is a land of soft colors of earth-gray sagebrush, lavender mountains, and brilliant colors of sky—vivid blue during the day, red and orange sunsets in the evening, and clear skies at night. The charm of the Southwest draws one close to the land; perhaps that is why one wishes to know the plants of this beautiful country.

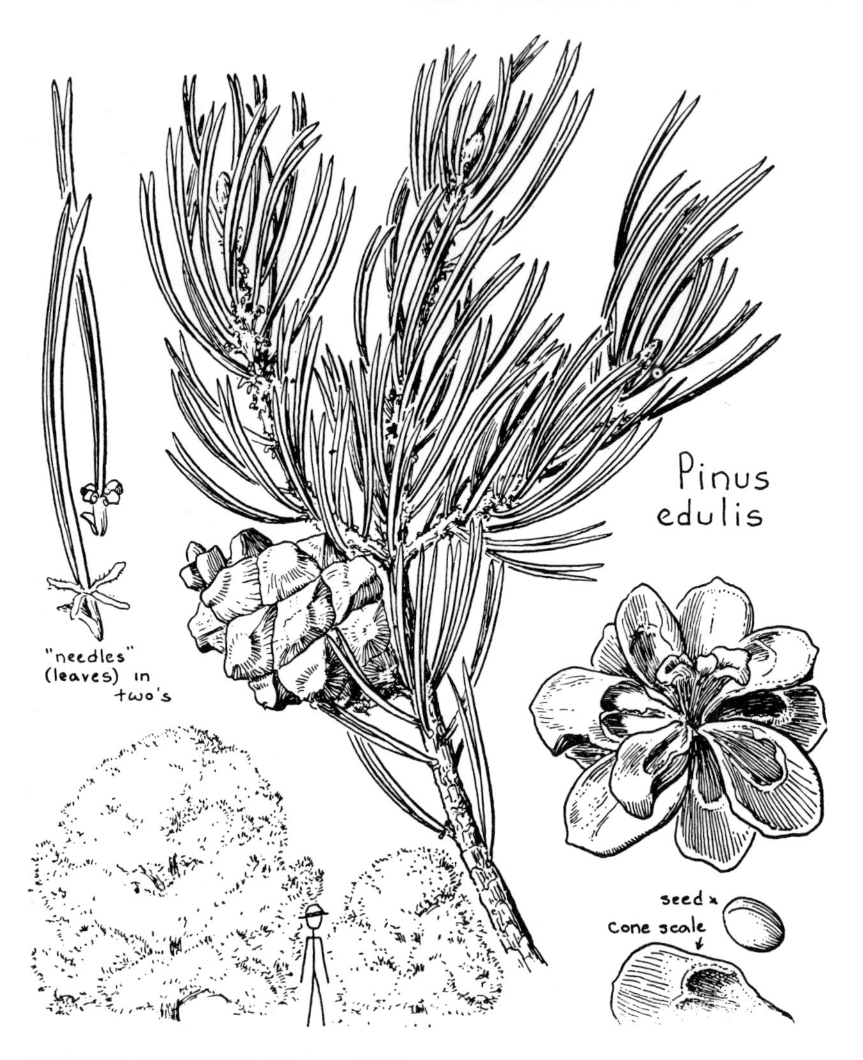

PINYON, NUT PINE (Pinus edulis)
Pine family *(Pinaceae)*
Colorado and Utah to Texas and northern Mexico, 5,000-7,000 feet.

The pinyon is a characteristic tree of the mesa country of the Southwest. It is small and scrubby, with a rounded crown and produces edible nuts borne in pockets at the base of the cone scales which spread open at maturity, dropping the nuts.

The needles are about 2 inches long, dark green, somewhat curved and are arranged in bundles of two or sometimes occur singly. The old bark is yellowish or reddish brown, irregularly furrowed and broken into superficial, small scales.

The nut is a source of food for Indians of the Southwest and is valued to a considerable extent by white man. Resin from the pinyon is used by Indians to waterproof baskets and pots and to cement turquoise stones in their jewelry. Pinyon wood, although soft, makes good fuel and gives off a fragrant smoke.

Especially fine stands of pinyon occur at Mesa Verde and in the Grand Canyon region. The singleleaf pinyon *(Pinus monophylla)* closely resembles the pinyon here described.

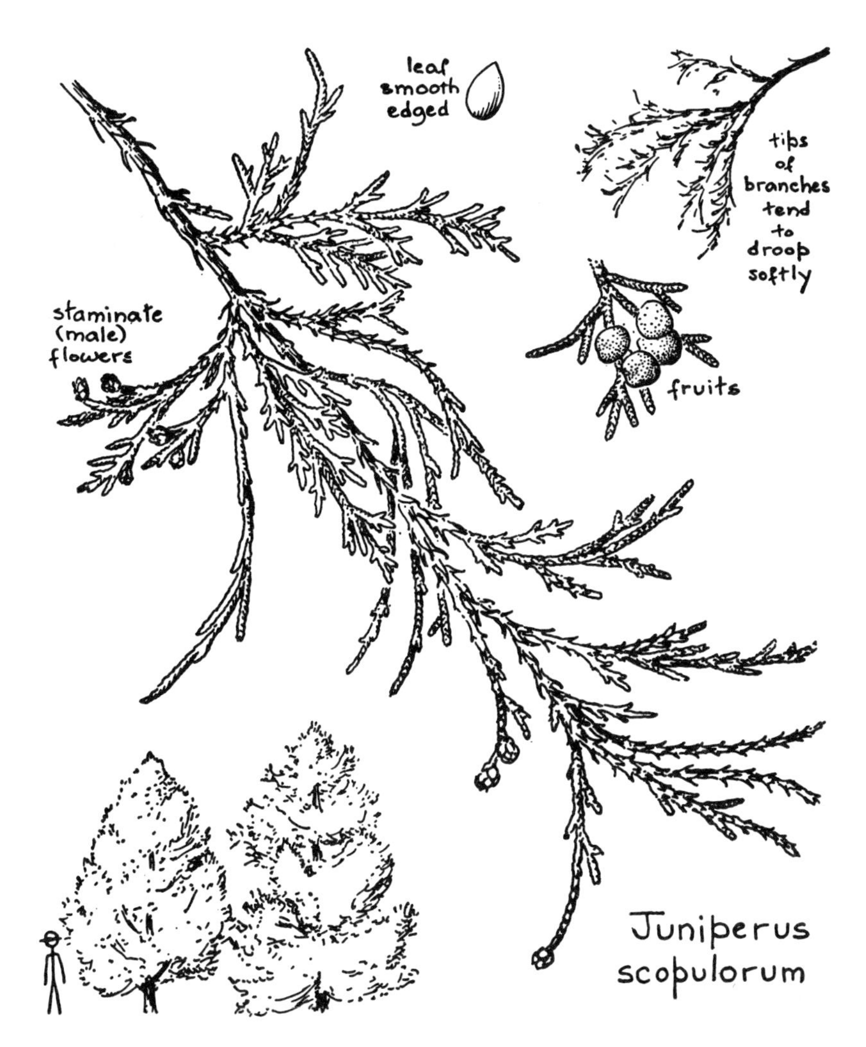

ROCKY MOUNTAIN JUNIPER, COLORADO JUNIPER
(Juniperus scopulorum)
Cypress family *(Cupressaceae)*
Alberta and British Columbia to New Mexico, Arizona and Nevada. Mesas, foothills and mountain sides, 5,000-9,000 feet.

This widely distributed juniper grows in the higher elevations of the Pinyon-Juniper Belt and on up into the Ponderosa Pine Belt. It is our most graceful and ornamental juniper, although the foliage is rather sparse. Growing about 20 feet high, it is conical in shape with silvery, drooping, flattened branches which give it the name "weeping juniper."

The small, succulent berries are blue and have a delicate, powdery surface coating. The trunk is short, as branching is close to the ground, and is about 18 inches in diameter. The bark is fibrous. The unseasoned heartwood is reddish or purplish.

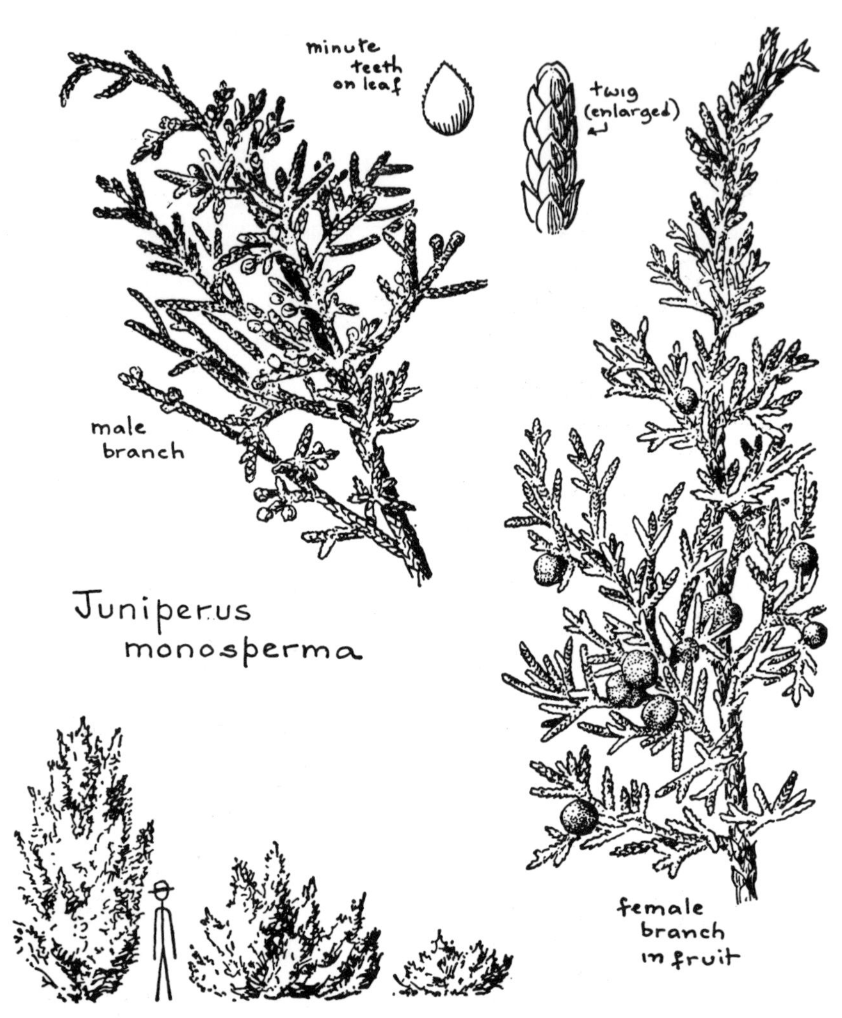

ONE-SEED JUNIPER, CHERRYSTONE (Juniperus monosperma)
Cypress family *(Cupressaceae)*
Colorado to Nevada into northern Arizona and New Mexico, 3,000 to 7,000 feet.

This common scrub tree is abundant in New Mexico, growing with, and lower than, the pinyon. The yellowish green foliage is bunchy. Leaves are tiny scales; the cone berry-like, small and bluish or copper colored. Male and female flowers grow on different plants, which explains why only some of the trees have berries. The tree branches close to the ground so that the trunk is not evident.

A similar appearing tree is the Utah juniper. *(Juniperus osteosperma)*, abundant in Utah and northern Arizona. In contrast, the foliage of this tree is not bunchy. The trunk is evident, often gnarled and twisted. The brownish berry is larger and less succulent than that of the one-seed juniper, and the male and female flowers grow on the same tree. Both species have a shreddy bark.

Juniper berries provide food for birds and wild animals, and are used in the making of gin and oil of juniper. The wood, being hard, is valuable for fence posts and fuel. The Hopi use juniper for medicinal purposes and in rituals.

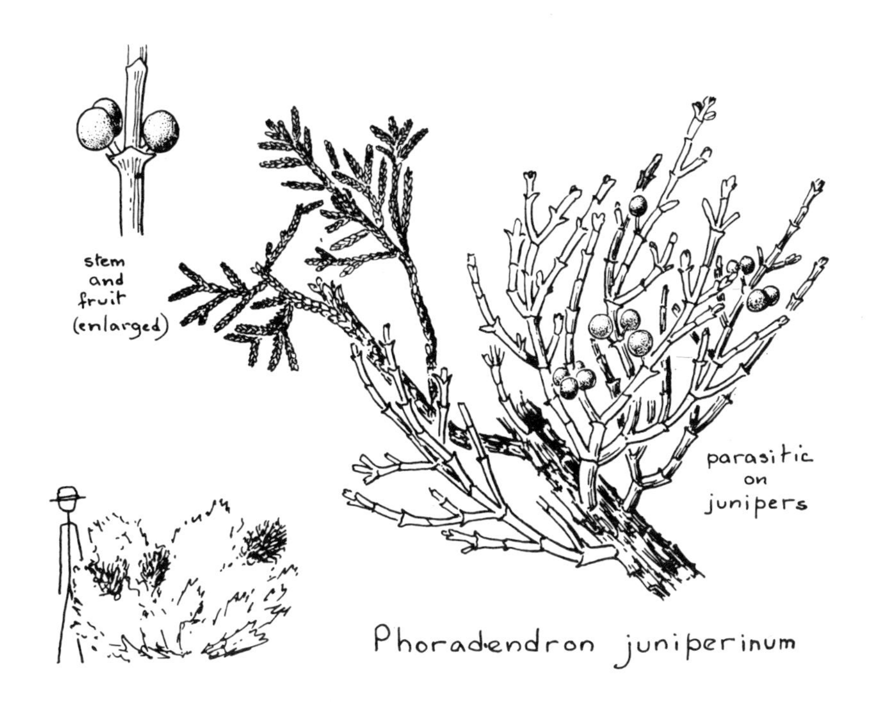

JUNIPER MISTLETOE (*Phoradendron juniperinum*)
Mistletoe family *(Loranthaceae)*. Blooms green.

Southwest Colorado and southern Utah to west Texas, Arizona, and northern Mexico, 4,200-7,000 feet.

This plant is a parasite on several species of juniper. Growing in bunches on the juniper branches and sapping the strength of the tree, it can be very harmful if it occurs in abundance.

The stout, yellowish-green, jointed stems are crowded and sometimes a foot long. They are brittle when dry. The leaves are reduced to triangular scales.

The greenish and inconspicuous male and female flowers grow on different plants. The pearl-like berries, partially transparent and whitish or pinkish, are relished by birds which carry the sticky seeds from tree to tree, thus spreading the mistletoe.

Hopi Indians use this species of mistletoe medicinally. The Papago dry the berries of one species and store them for winter food. The berries of some species are poisonous.

This mistletoe is not as decorative as those with larger leaves, which are gathered in large quantities and shipped east for decorations at Christmas time.

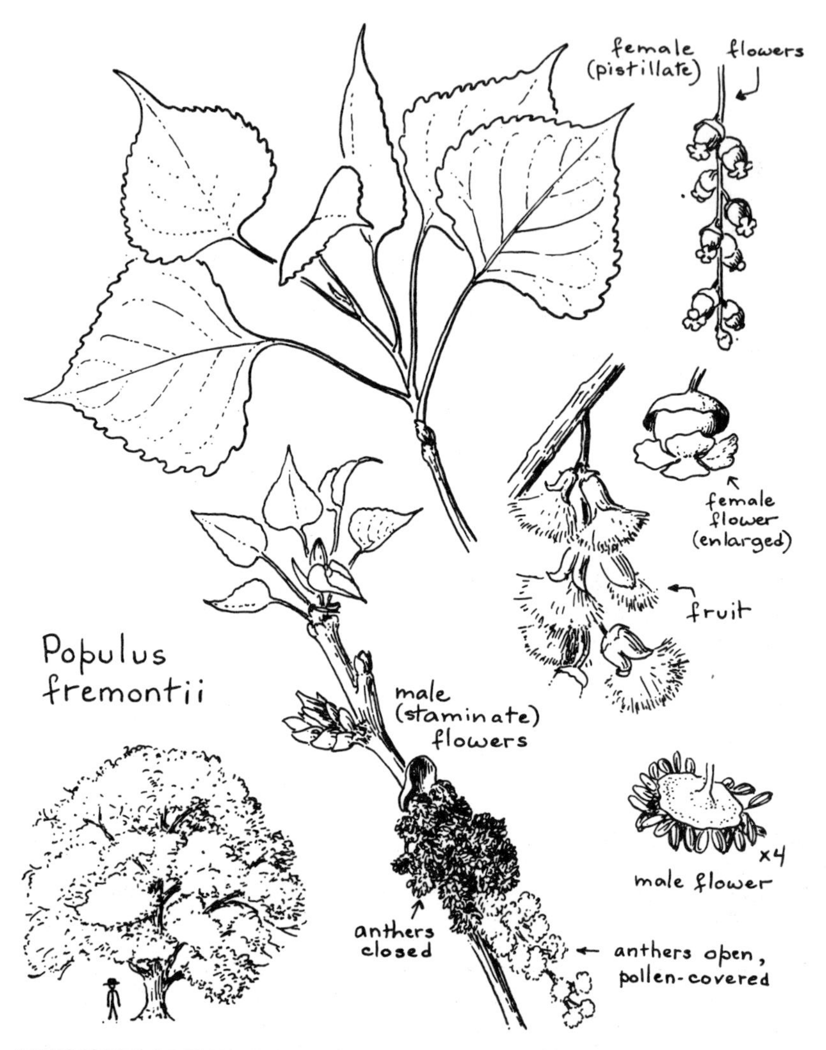

FREMONT COTTONWOOD (Populus fremontii)
Willow family *(Salicaceae)*
West Texas to Nevada, Arizona, California and northern Mexico, 6,000 feet or lower.

One of the few deciduous trees of the semi-desert is the cottonwood—a strictly southwestern species—growing along streams and in other moist places. It is a conspicuous tree, reaching a height of from 50 to 100 feet. The trunk, sometimes as much as 4 feet in diameter, has thick, deeply furrowed bark of a gray-brown color. The main branches are large and yellowish in color while the wide crown is flat-topped.

Flowers grow in drooping catkins, blooming early in the spring before the leaves appear. Only the female trees produce seeds which are covered with silky or cottony hairs.

This rapid growing, short-lived cottonwood provides welcome shade in a country of brilliant sunshine, but it is valuable for fuel, fence posts and building material only when there is no better wood available.

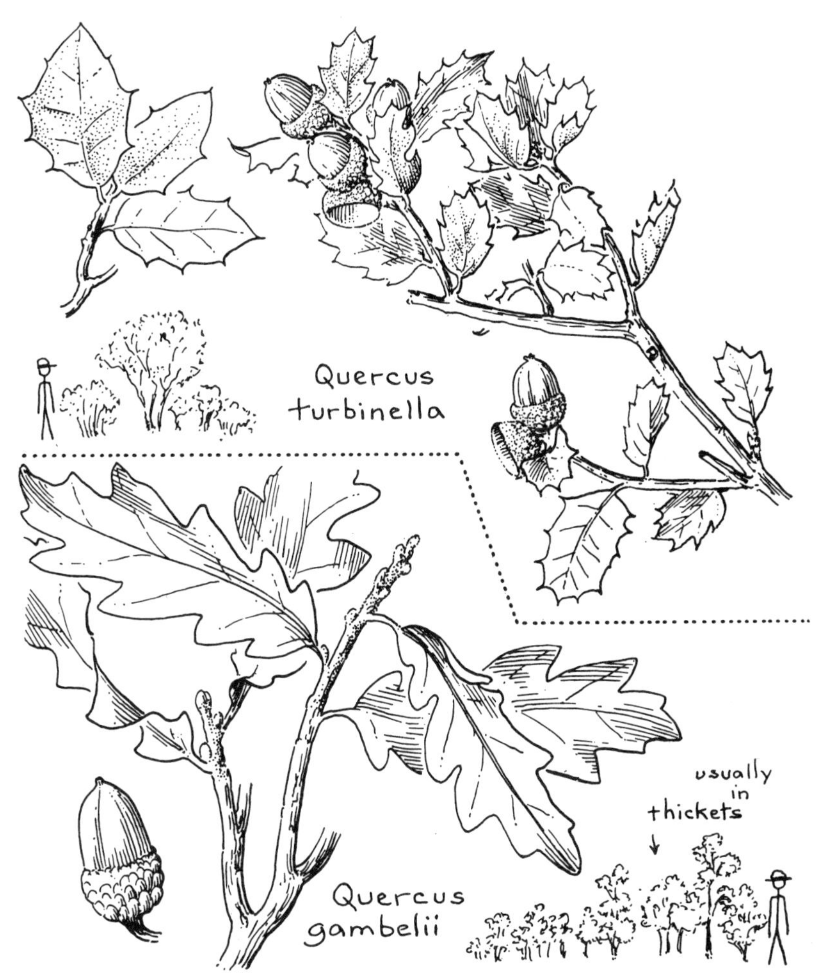

GAMBEL OAK, ROCKY MOUNTAIN WHITE OAK (Quercus gambelii)
Beech family *(Fagaceae)*
Colorado to Nevada, south to northern Mexico, 5,000-8,000 feet.

This shrub or tree, from 6 to 50 feet high, forms thickets in the Pinyon Belt and also grows under the ponderosa pines. The bark is gray, scaly or flaking and yields a tanning material.

The deeply lobed, deciduous leaves are browsed by deer and livestock. As in all oaks, male and female flowers grow on the same tree. The fruit, an acorn half covered by its cup, is fed upon by birds, squirrels and other wild animals. Acorns are used in many ways by Indians, and form an important part of the browse of deer.

SHRUB LIVE OAK (Quercus turbinella)
Colorado to New Mexico, to California and northern Mexico, 8,000 feet, commonly lower.

An evergreen shrub seldom more than 13 feet high, it forms a chaparral in the Pinyon Belt and is valuable in retarding soil erosion and as a winter feed for cattle. The leaves have spiny teeth, are bluish green above, tawny below and covered with a fine, white powder. The acorn cup is top-shaped.

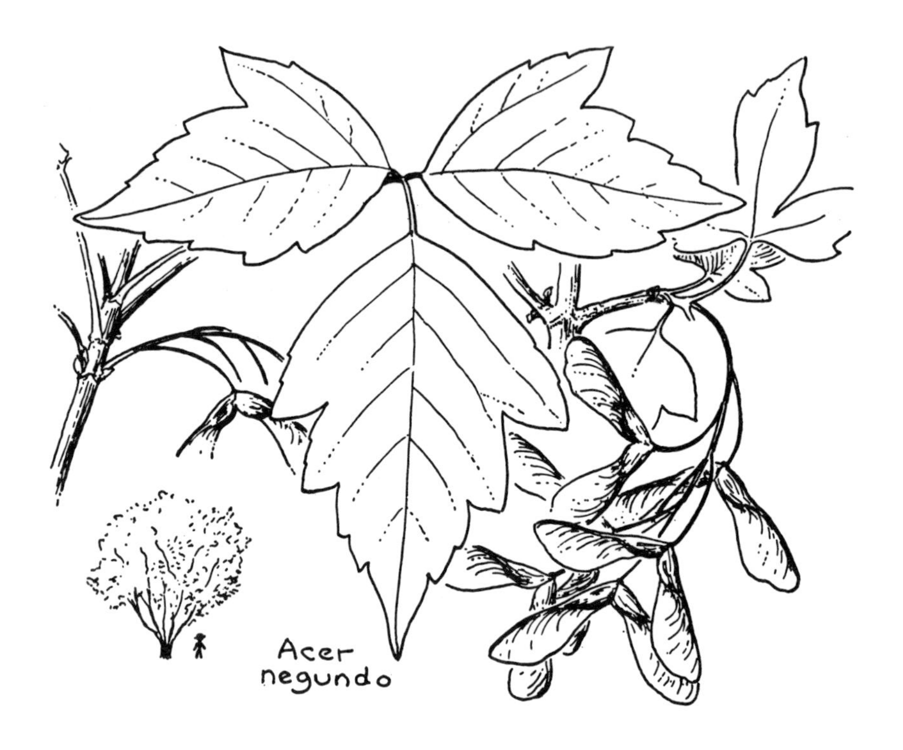

BOXELDER, ASH-LEAVED MAPLE (Acer negundo)
Maple family *(Aceraceae)*. Blooms greenish, April.
Through the Southwest and northward to Canada, 5,000-8,000 feet.

The boxelder grows along streams of the Pinyon and Ponderosa Pine Belts; for example, there are many fine specimens on the floor of Frijoles Canyon at Bandelier National Monument, and along the Virgin River in Zion Canyon. It is also found at Mesa Verde, and in Walnut Canyon, Montezuma Castle, Navajo and Canyon de Chelly National Monuments.

A rapid-growing, freely branched tree about 50 feet high, it has a short trunk, broad crown, and light green twigs. The bark is pale, grayish brown.

The leaves are compound with three to five leaflets. The blades are rather thick and slightly hairy. They grow opposite each other on the branches, as do all maple leaves.

The flowers are greenish, arranged in drooping clusters, and bloom in April before the leaves appear. Male and female flowers grow on different trees. The fruit has two broad, finely veined wings.

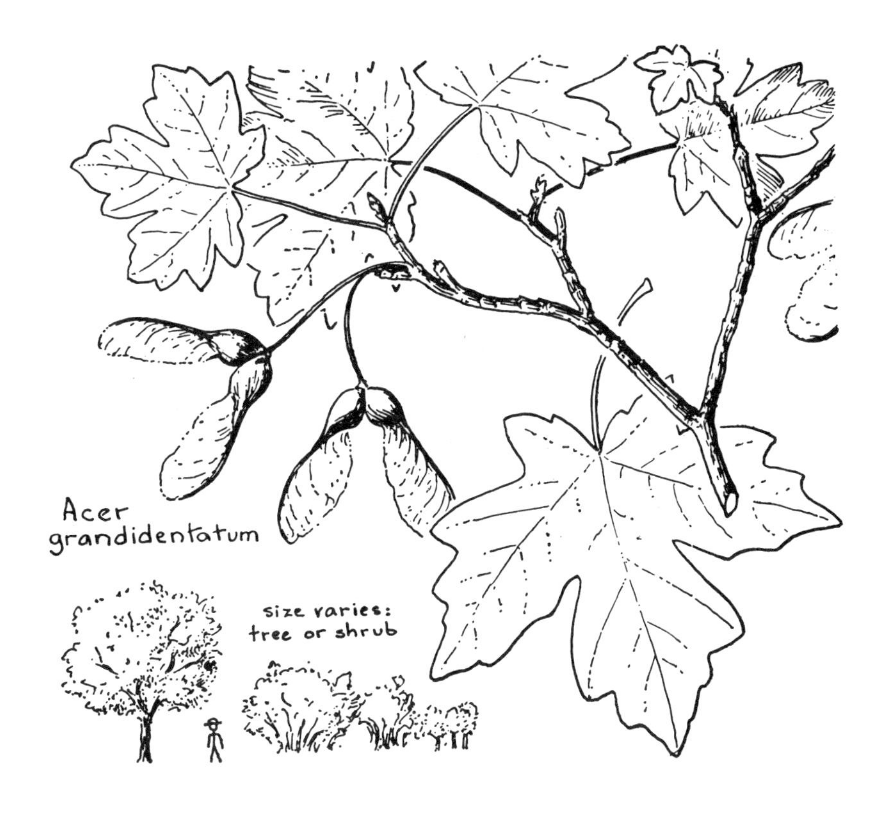

BIGTOOTH MAPLE, SCRUB MAPLE (Acer grandidentatum)
Maple family *(Aceraceae)*
Montana and Idaho to west Texas, New Mexico and Arizona, 4,700-8,000 feet.

In the fall of the year one of the beautiful sights in the Southwest is this scrub maple, making canyon walls and mountain sides brilliant with splashes of red. Its bright colors are conspicuous in the upper part of Zion Canyon, on the rim of Bryce Canyon, at Mesa Verde and at Grand Canyon as well as Bandelier National Monument and in the mountains near Santa Fe.

This shrub or small tree, sometimes 50 feet high, inhabits damp places of the Pinyon, Ponderosa and Aspen Belts. The trunk is about 1 foot in diameter, the bark smooth, gray or brownish, the sap sweet. The wood makes excellent fuel.

Leaves are thickish, slightly hairy beneath, and are browsed by deer and livestock.

Flowers of the maples, with the exception of the boxelders, are called "polygamous" because male and female parts are sometimes in the same flower, and sometimes in different flowers either on the same or on different plants.

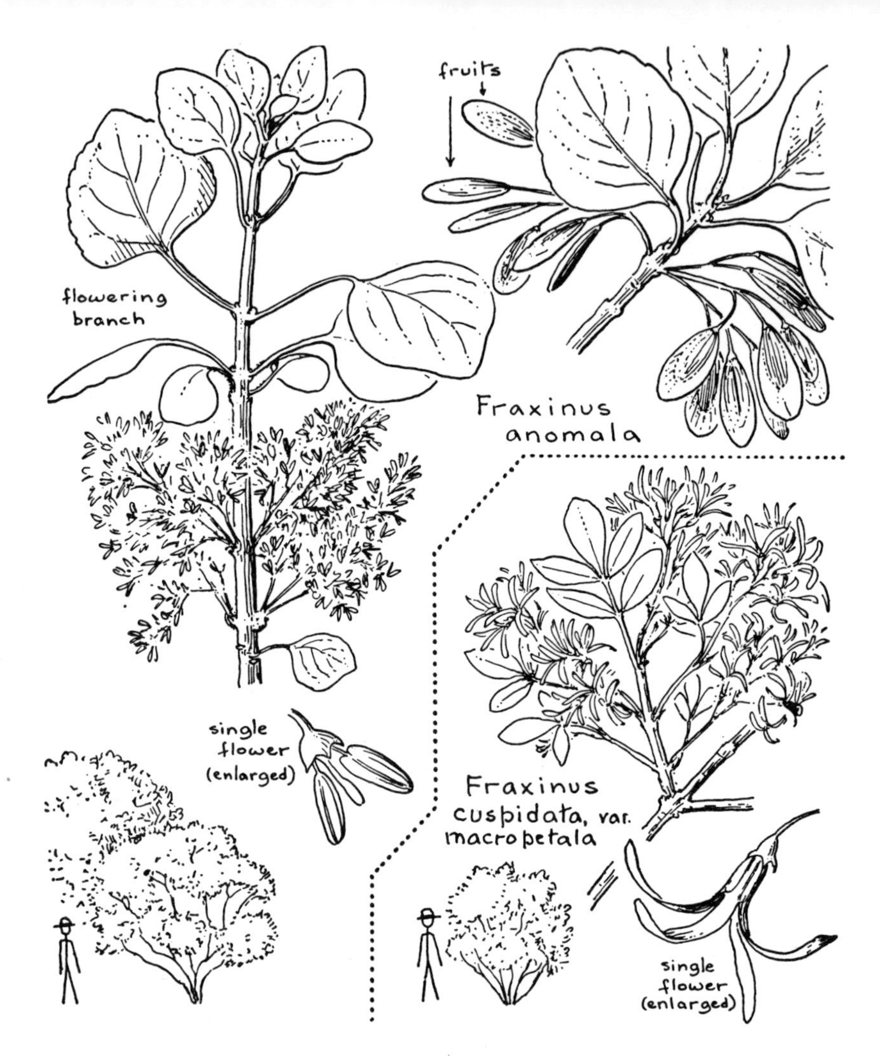

SINGLELEAF ASH (Fraxinus anomala)
Olive family *(Oleaceae)*. Blooms orange and green, April.
Western Colorado and New Mexico to northern Arizona and southeastern California. 2,000-6,000 feet.

The singleleaf ash is a shrub with dark brown bark slightly tinged with red. The leaves are smooth and dark above, slightly paler below and are thin but rather leathery. The anthers of the flowers are orange.

This shrub is a relative of the olive and also of the garden lilacs and privet. It grows on dry hills and in canyons and is the principal deciduous tree of the pinyon-juniper forest.

It is especially common in the Grand Canyon and in Aztec Ruins, Canyon de Chelly and Arches National Monuments.

FLOWERING ASH (Fraxinus cuspidata var. macropetala)
Same family.
Blooms white, May-June. Arizona.

The flowering ash grows about as large as a lilac bush. The drooping plumes of white flowers are fragrant; petals are long and narrow and look like a white fringe.

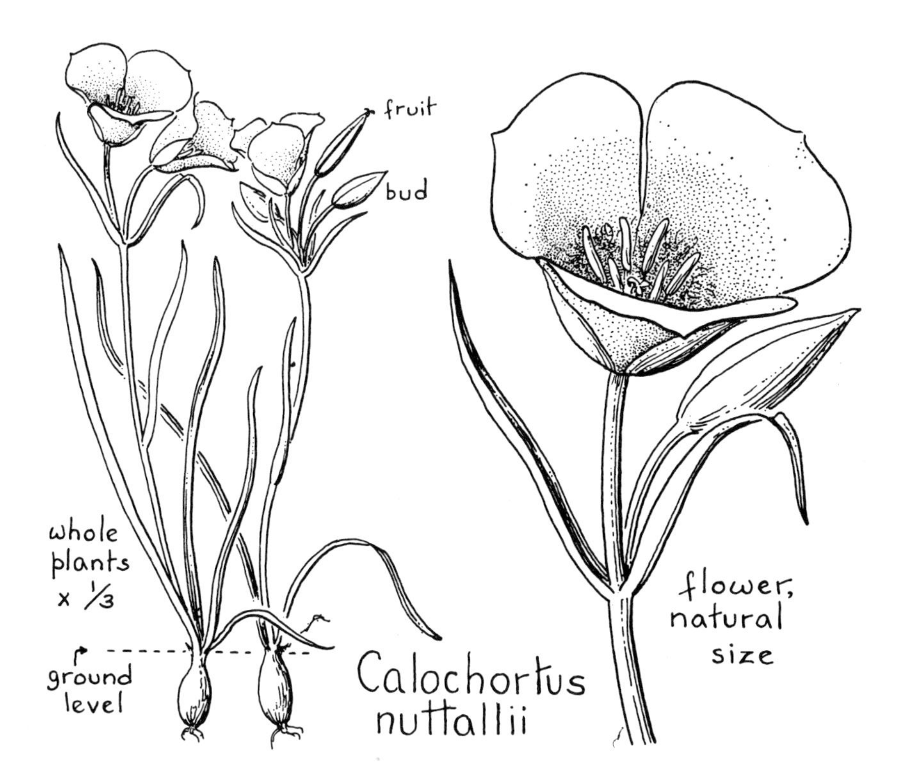

SEGO-LILY, MARIPOSA (Calochortus nuttallii)

Lily family *(Liliaceae)*. Blooms whitish to lavender-blue, May-July.

Western North Dakota to Eastern Oregon, south to Nebraska, northern New Mexico and Arizona, and eastern California, 5,000-8,000 feet.

One of the most beautiful wild flowers of the Southwest is the sego-lily mariposa, the state flower of Utah. This graceful, tulip-like flower varies in color from cream-white tinged with purple to lilac or blue. It is well named calochortus, for in Greek this means beautiful herb.

It associates with sagebrush on dry mesas and foothills and also in open mountain forests, and is found at Grand Canyon, Zion Canyon and Mesa Verde National Parks, and at Aztec Ruins and Montezuma Castle National Monuments.

The cup-shaped flower is composed of three pointed green sepals, and three petals beautifully marked at their bases, first with yellow near the center, then a purplish spot, and farther out a variously fringed or bordered gland.

There are from one to five flowers on a slender stalk; and from one to several grass-like leaves which often wither before the flowers bloom. The flower stalk rises about 1 foot from an onion-like bulb.

The bulbs are eaten by the Hopi and Navajo, and the early Mormon pioneers of Utah used them for food in time of scarcity.

The golden sego-lily (var. *aureus*) is similar to the above except that it is a shorter plant and the petals are lemon-yellow. This variety occurs in the petrified forests in Zion and Petrified Forest National Parks, and at Canyon de Chelly and Navajo National Monuments.

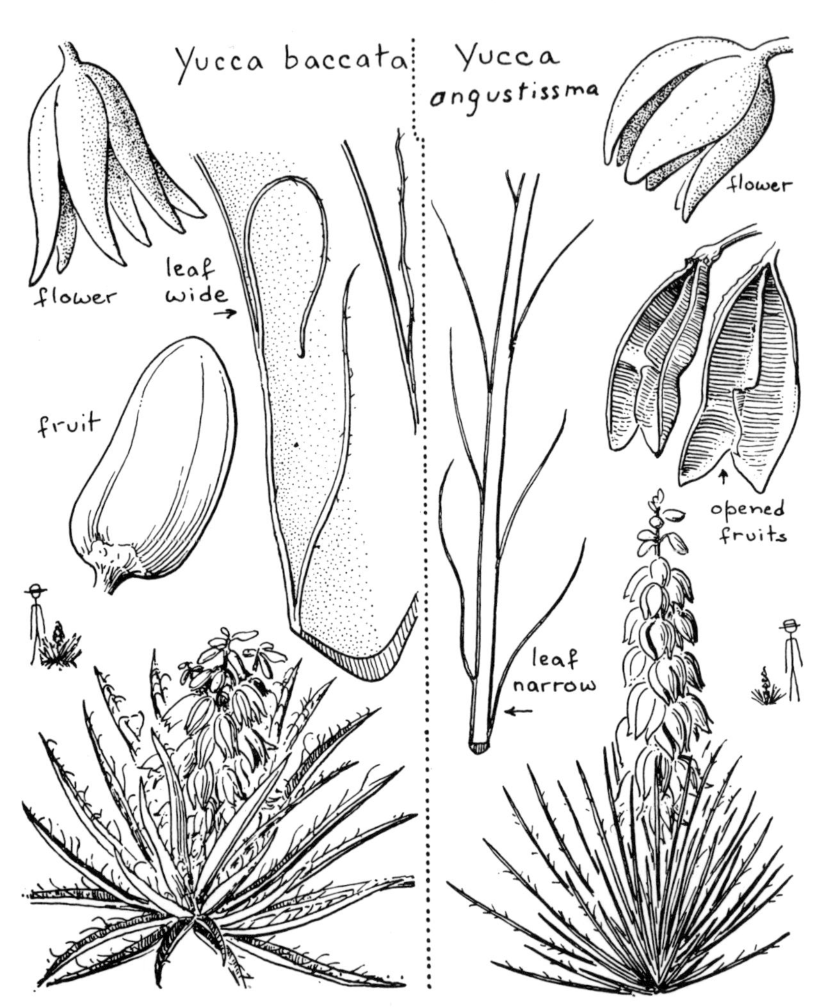

YUCCA, SPANISH-BAYONET, SOAPWEED YUCCA
Lily family *(Liliaceae)*. Blooms white, tinged with purple or brown; bell-shaped.

The yucca is pollinated by a small moth whose larvae feed on the seeds. There are ragged looking fibers along the margins of the spine-tipped leaves. Indians avail themselves of the buds, flowers, fruits, seeds and young flower stalks for food and a fermented beverage is made from the fruits. Its fiber as well as its leaves are used for baskets, mats, cloth, rope and sandals, while soap is made from the roots.

DATIL YUCCA, BANANA YUCCA, (Yucca baccata)
With broad stiff leaves. Blooms April-July.
Southern Colorado to Nevada, western Texas, Arizona and southern California, 3,000-8,000 feet.

FINELEAF YUCCA (Yucca angustissima)
Blooms May-June.
Southern Utah and Nevada, northern New Mexico and Arizona, 2,700-7,500 feet.

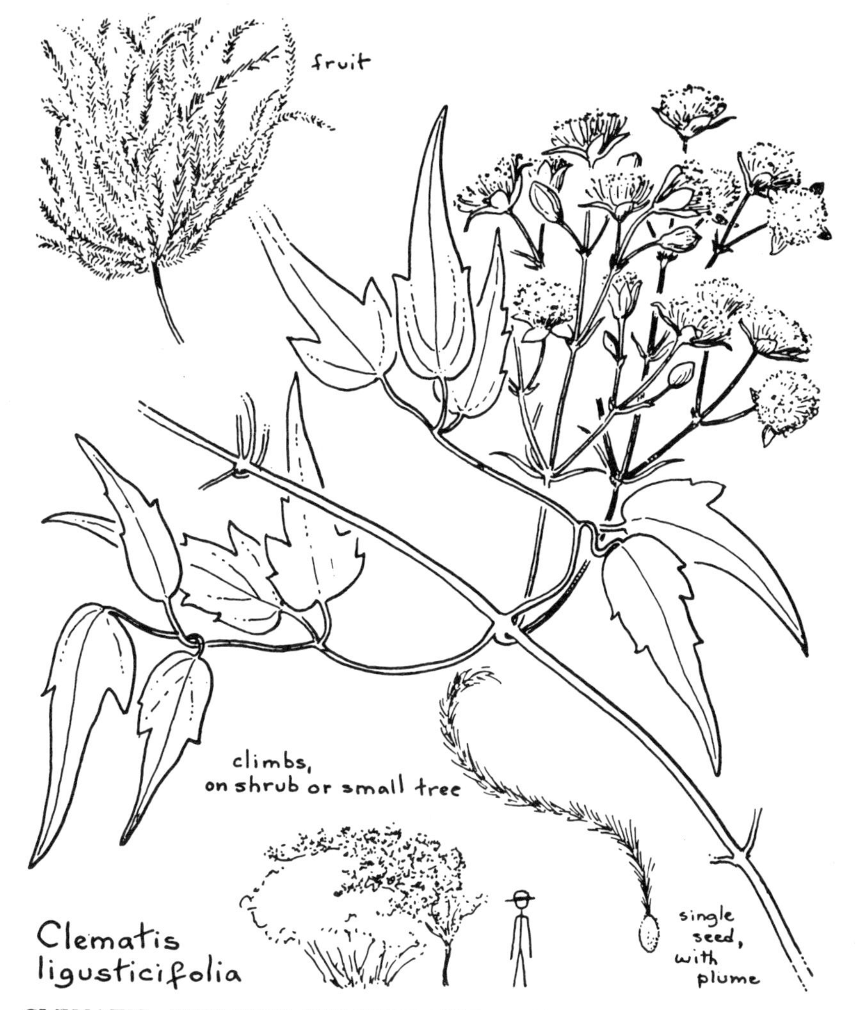

CLEMATIS, WESTERN VIRGINSBOWER (Clematis ligusticifolia)
Buttercup family *(Ranunculaceae)*. Blooms white, May-September.
Western Canada and North Dakota to New Mexico, Arizona and California, 4,000-7,000 feet.

Clematis is a slender trailing vine that makes lovely, light green bowers by growing over rocks and the bushes of the chaparral. The stem has a woody base and the long twisting leaf stems form the tendrils by which the vine clings.

Leaves are compound and opposite and the five leaflets grow some distance apart and look like separate leaves.

The flowers are white and grow in branched clusters; there are no petals, but the four sepals are petal-like. The male (staminate) flowers are the handsomest with their feathery stamens. The female (pistillate) flowers grow on a different plant and produce the beautiful plumed seeds.

The Indians used clematis as a remedy for sore throats and colds. It is said that the crushed roots of clematis were sometimes placed in nostrils of tired horses to revive them.

This species of clematis grows in Grand Canyon, Zion and Mesa Verde National Parks and in Montezuma Castle, Walnut Canyon, Aztec Ruins, Chaco Canyon, Navajo and Bandelier National Monuments.

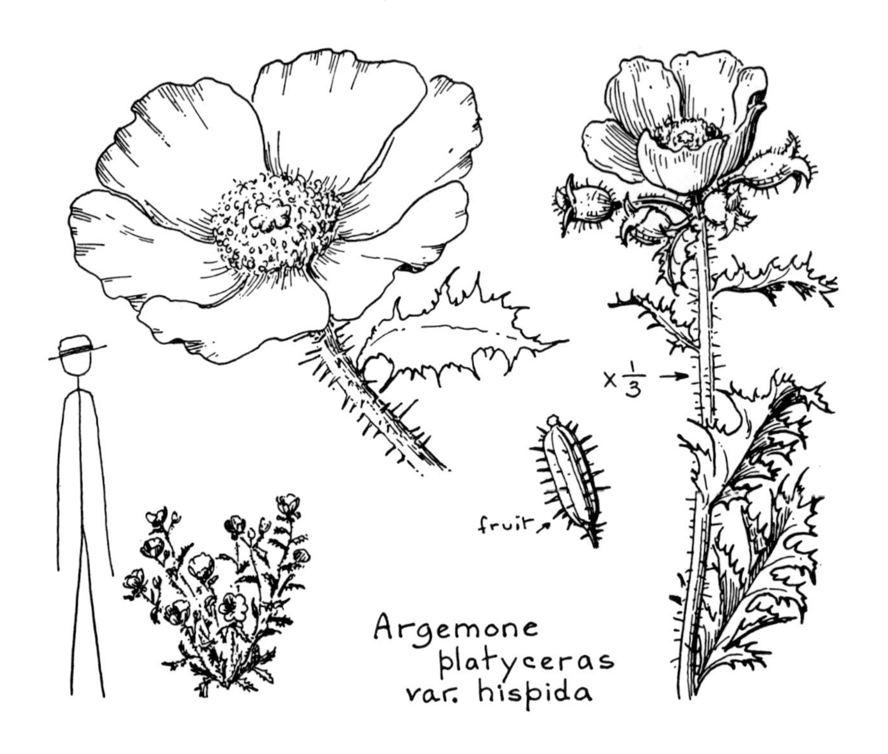

THISTLE POPPY, PRICKLE-POPPY (Argemone platyceras var. hispida)
Poppy family *(Papaveraceae)*. Blooms white, flowering almost throughout the year.
Nebraska and Wyoming to Arizona and Mexico, 1,400-8,000 feet.

Thistle poppy, with its coarse, prickly stems and leaves, and its large, delicate, white flowers, is a typically western plant of the open cattle country. Decidedly drought-resistant, it has a wide climatic range, frequenting hot desert sands, arroyos, foothills, and open mountain parks. Its presence is often an indication of overgrazing.

This poppy is found at Grand Canyon, at House Rock Valley to the east of the Kaibab Plateau, in Zion Canyon, and in the sagebrush country around Aztec, New Mexico.

The flowers are conspicuous and large, from 2-5 inches across and there are several to a plant. The thin petals are separate, usually six or sometimes four in number, twice as many as the horn-tipped sepals. These sepals, usually three or sometimes two in number, have peculiar, short, thick horns which give a queer, horned appearance to the buds. In the center of the flower are many bright, orange-yellow stamens surrounding the short style which is capped with a purplish, star-like stigma. Seeds, many in a pod, are relished by doves.

The bitter sap of this poppy has been used to treat skin diseases. The scientific name, *Argemone,* comes from the Greek and means "a kind of poppy."

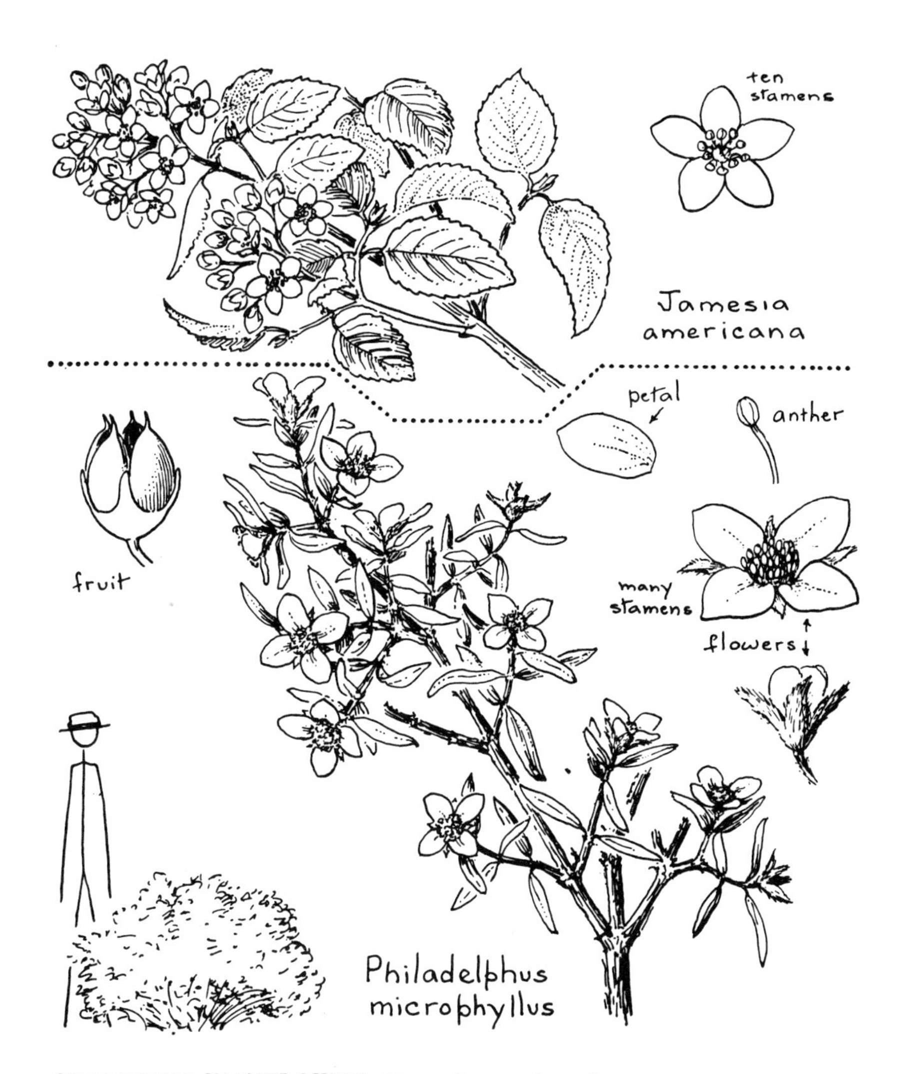

CLIFFBUSH, WAXFLOWER (Jamesia americana)
Saxifrage family *(Saxifragaceae)*. Blooms waxy white or pinkish, July.
Wyoming to New Mexico and southeastern Arizona, 7,500-9,500 feet.

Cliffbush occurs only in the upper reaches of the pinyon-juniper woodland. The brownish bark is shreddy; branches and twigs, reddish brown. The thin leaves are green above, downy white beneath and turn red in the fall.

MOCKORANGE, SYRINGA (Philadelphus microphyllus)
Same family. Blooms white, June and July.
Southern Colorado, New Mexico, Arizona and California, 5,000-8,000 feet.

The leaves are very numerous on the branches of this shrub. They are rather thick, bright green above, and have white silky down beneath. The bark flakes off in shreds.

The flowers are pretty and smell like lemon blossoms. The mockorange is browsed by mountain sheep.

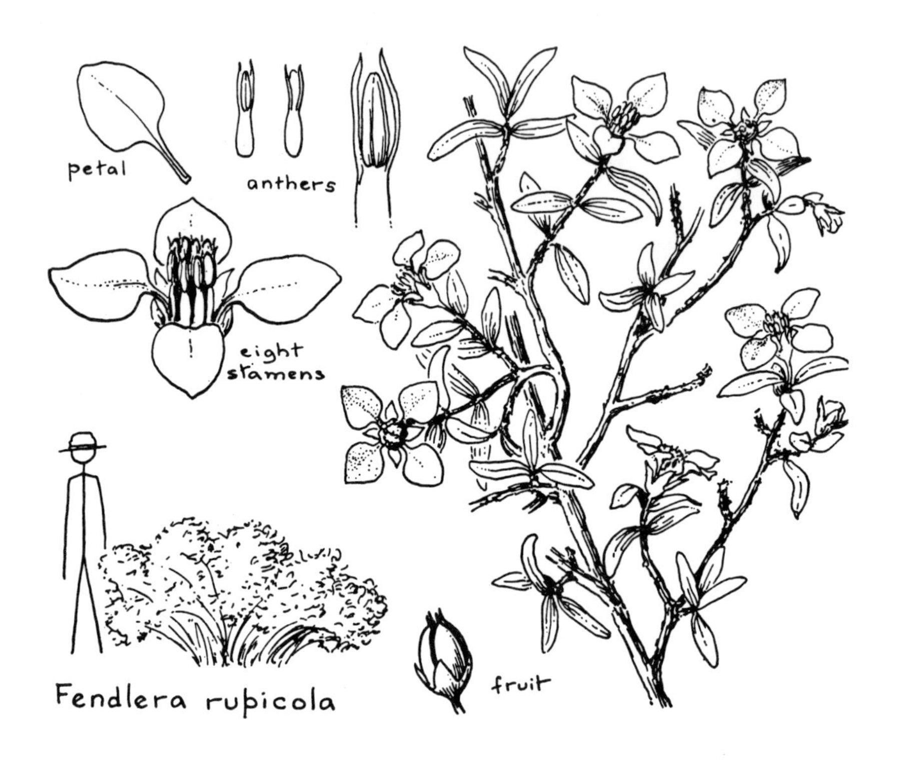

CLIFF-FENDLER BUSH (Fendlera rupicola)
Saxifrage family *(Saxifragaceae)*. Blooms white, March-June.
Southern Colorado to western Texas and Arizona, 4,000-7,000 feet.

A decorative shrub, straggling, intricately branched, growing about 3 feet high or more. It has furrowed gray bark. The leaves are green beneath as well as above and the leaf margins often curl under.

The fragrant flowers are large and showy and are white tinged with purple. This plant resembles the mockorange *(Philadelphus microphyllus)* but can easily be distinguished from it by the stem-like base of the petals and the eight stamens. The mockorange has many stamens.

Goats and deer browse it and it is used to some extent by cattle.

It grows on rocky, gravelly slopes in Grand Canyon and Mesa Verde National Parks and in Aztec Ruins, Chiricahua, and Navajo National Monuments.

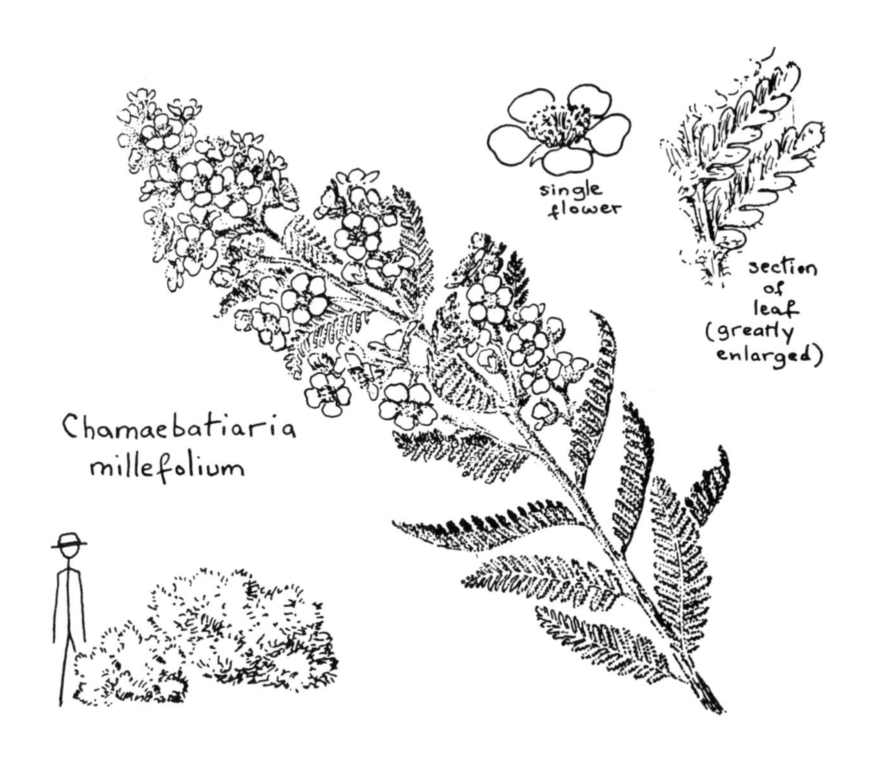

FERNBUSH, TANSYBUSH, DESERTSWEET (Chamaebatiaria millefolium)
Rose family *(Rosaceae)*. Blooms creamy-white, July-August.
Idaho to Arizona and California, 4,500-7,000 feet.

Fernbush is one of the most beautiful and ornamental shrubs that grow among the rocks with the sagebrush and pinyons. On both rims of Grand Canyon it is common and spectacular. Large symmetrical bushes grow around Yavapai Observation Station on the south rim of Grand Canyon, and it is very common in Walnut Canyon National Monument.

This aromatic shrub is full and rounded, branching profusely. The stems are reddish, the bark shreddy. The thick, leathery leaves are more or less evergreen, are many times divided and look a great deal like fern leaves. The leaf surface is slightly scaly and sticky. The plant is browsed by sheep, goats and deer, but apparently not by cattle.

In mid-summer this bush is heavy with bloom. Although the blossoms are small, they are so numerous that they are very showy. They resemble a strawberry blossom or a small wild rose, and grow in full elongated clusters. They are slightly fragrant. Fruits are dry, simple pods.

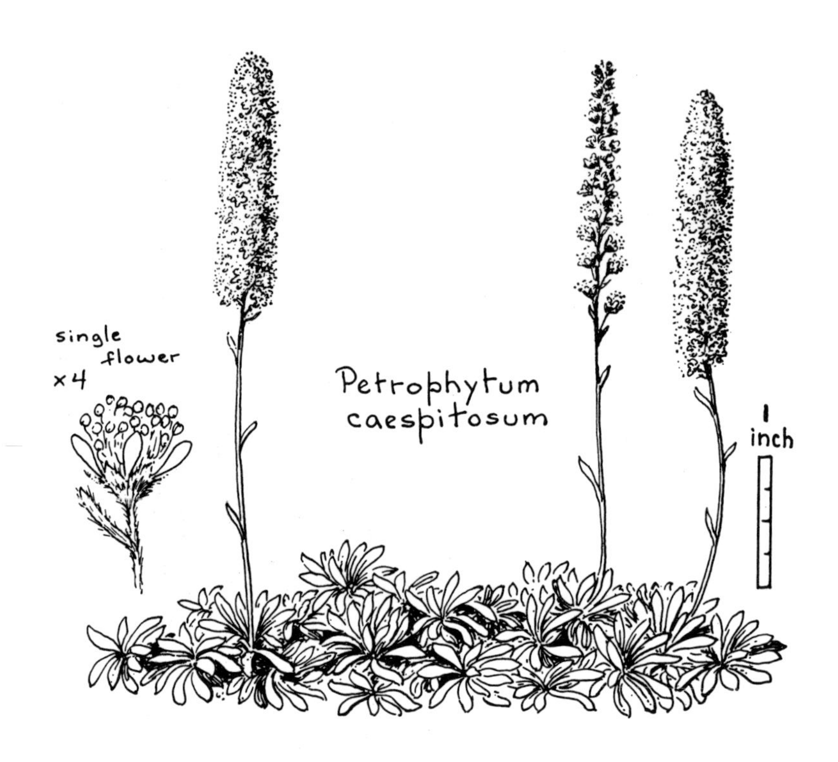

ROCK SPIRAEA, ROCKMAT, "ROCKROSE" (Petrophytum caespitosum)

Rose family *(Rosaceae)*. Blooms white, July-September.

South Dakota and Montana to New Mexico, Arizona and California, 5,000-8,000 feet.

The rock spirea forms pretty gray-green mats over dry rocks, and sends up spikes of feathery white flowers. It is an undershrub; the branches lie close to the rocks, the roots finding their way into crevices.

The calyx and leaves are covered with silky hairs; leaves are arranged in closely spaced rosettes.

The small flowers in dense spikes have many stamens which are longer than the rather inconspicuous petals and give the spike a fluffy appearance.

It is not a common plant but is found on both rims of Grand Canyon where it creeps over the limestone ledges that look too dry to support any kind of life. In Zion Canyon it may be seen from the Great Arch Trail and near the pueblo food cache opposite the Great Organ. It also grows in Walnut Canyon and Navajo National Monuments.

It is well named petrophytum, for in Greek, *petra* means rock and *phyton* means plant.

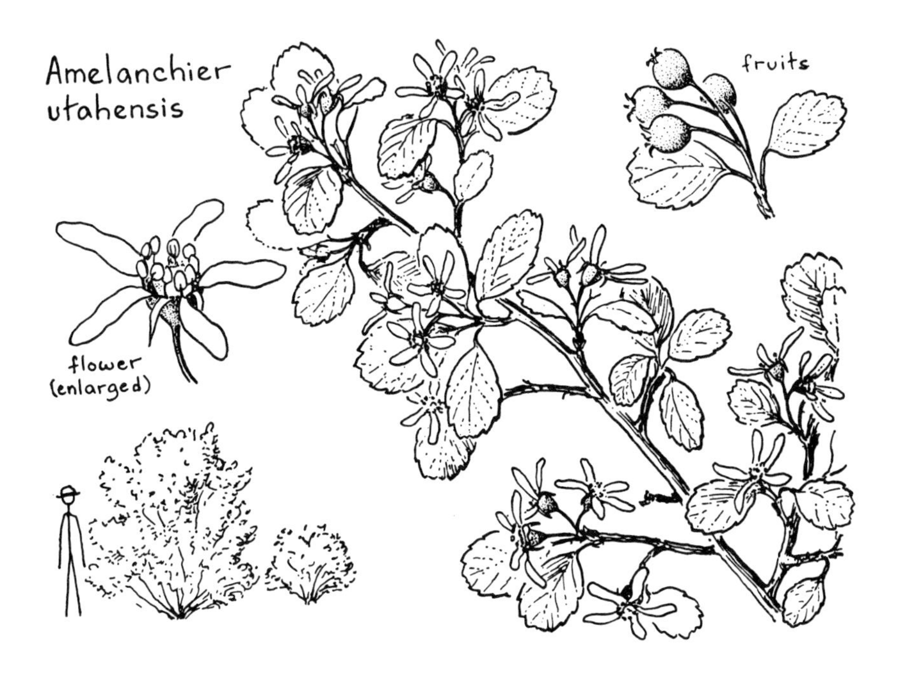

SERVICEBERRY, SHADBUSH (Amelanchier utahensis)
Rose family *(Rosaceae)*. Blooms white, May.
Colorado to Nevada, New Mexico and northern Arizona, 2,000-7,000 feet.

Serviceberry is a shrub or small tree growing on dry hills and mesas with sagebrush and pinyon. It is conspicuous in the spring, since it puts out flowers and leaves in advance of most of the other shrubs. It is common in and around Grand Canyon, at Zion Canyon, Mesa Verde, and in arroyos and among hills along the San Juan River, New Mexico.

The white flowers of the serviceberry occur in clusters. They are five-petaled and rest in the green bell-shaped tube formed by the five united sepals. There are many yellow stamens inserted on the rim of the flower tube.

The apple-like fruits ripen dry and yellow. They are insipid to taste; however, in earlier days, the Indians ate them fresh or dried and preserved them for winter use.

The simple leaves are nearly round, slightly narrowed at the tip, and finely and bluntly toothed. They are lighter colored and hairy on the under side. Twigs are brown.

The fruit and leaves of serviceberry may become infested with three kinds of rust whose spores infect the Utah juniper and the Rocky Mountain juniper. After doing some damage to the leaves and twigs of these trees, another crop of spores is produced which re-infect the serviceberries.

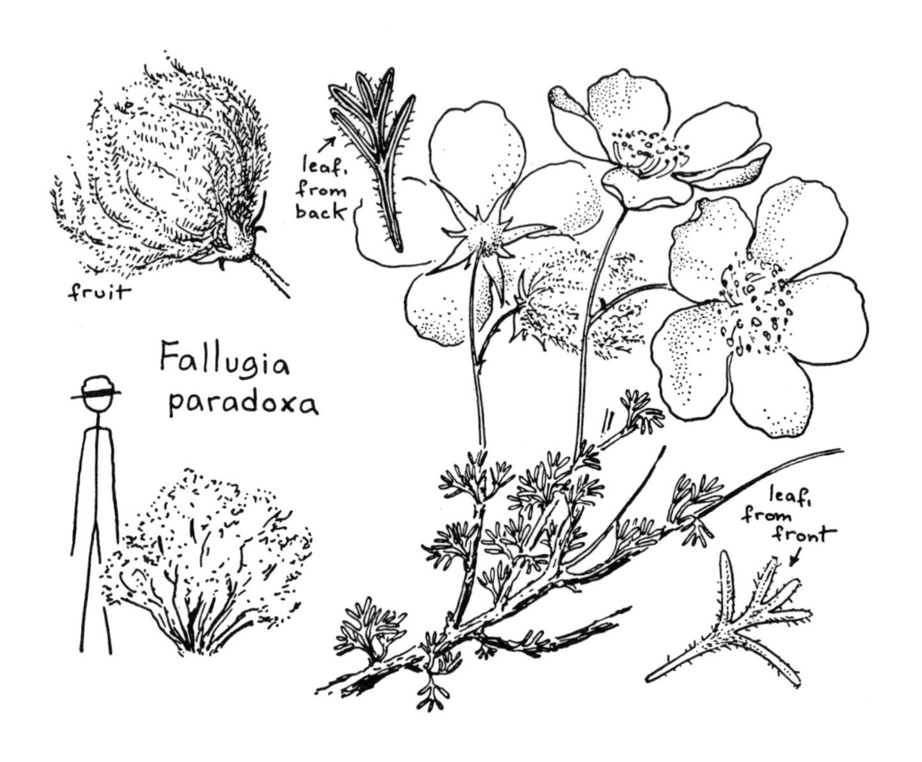

APACHE-PLUME (Fallugia paradoxa)
Rose family *(Rosaceae)*. Blooms white, April-October.
Southern Colorado, Utah, Arizona and New Mexico, 3,700-8,000 feet

Common in the southwestern chaparral of dry mesas and arroyos is the Apache-plume. It is a rounded, straggling shrub with many slender, woody, white branches, growing about 4 or 5 feet high. It is valuable as a soil binder and a browse for cattle, sheep and goats.

The white, yellow-centered flowers are large when there has been plenty of rain, but quite small if the season is dry. They are usually about the size of an apple blossom, have five broad petals and grow on long, slender stalks. The blooms are more scattered on the Apache-plume than on the cliffrose and therefore not as showy.

The pink or purplish plumes on the seeds grow in fluffy balls which sometimes cover the whole plant. They remain on the shrub for a long time and are fully as beautiful as the flowers.

The slightly downy, evergreen leaves are three to seven cleft. They somewhat resemble the leaves of cliffrose in shape, but do not have an odor. The Hopi steep the leaves to make a hair tonic.

Apache-plume is common in the Grand Canyon region, in the area around Santa Fe, and at Wupatki and Navajo National Monuments.

There is an unusually large stand of Apache Plume near the Rio Grande in Bandelier National Monument.

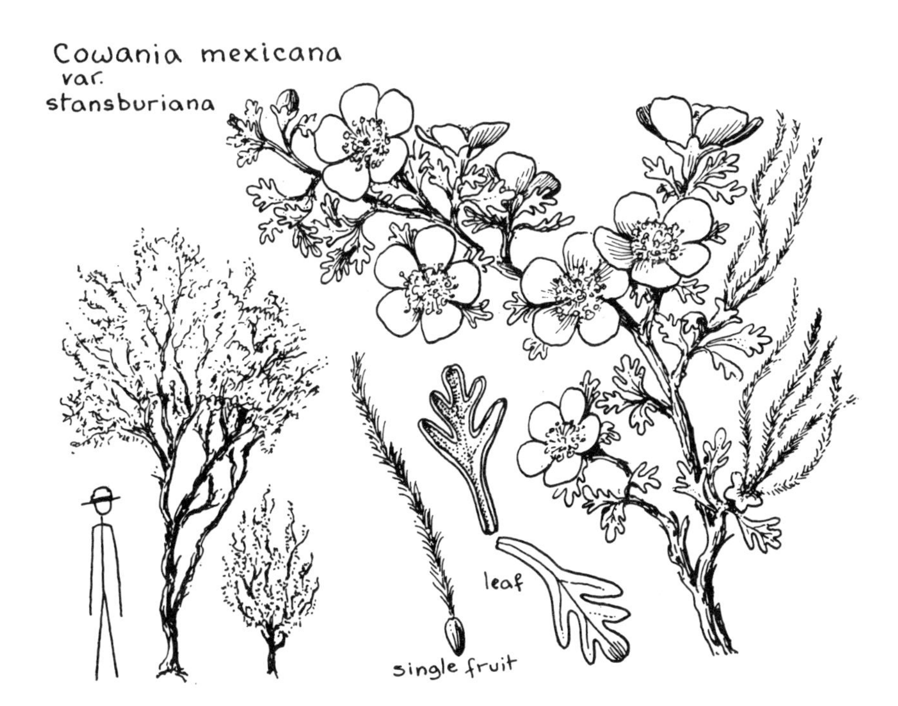

CLIFFROSE, BUCKBRUSH, QUININE BUSH (Cowania mexicana)
—See Cover

Rose family *(Rosaceae)*. Blooms creamy white or pale yellow, spring and summer.

Southern Colorado to Nevada, Arizona, California and northern Mexico, 3,500-8,000 feet.

One of the most attractive shrubs in this part of the Southwest is the cliffrose. It is a shaggy, gnarled and twisted bush, but is nonetheless beautiful, with creamy flowers growing thickly on the stems, and long white plumes on the seeds.

The five-petaled flowers with golden centers resemble small wild roses, but their fragrance is more like orange blossoms.

Although bitter to taste, cliffrose is one of the most important winter browse plants for cattle, sheep and deer.

The light gray, shreddy bark was used by early Indians in the making of clothing, sandals, mats and rope. Today the Hopi Indians use cliffrose as an emetic, and as a wash for wounds. The wood was formerly used in the making of arrows.

This shrub or small tree, from 4 to 25 feet high, grows on dry slopes and mesas with the pinyon and juniper trees.

A similar appearing, and very common shrub, also in the rose family, is the antelopebrush or bitterbush *(Purshia tridentata)*. It is more intricately branched than cliffrose and has small, yellow, inconspicuous flowers. The seeds are not plumed.

Antelopebrush grows from 4,000 to 9,000 feet and blooms from April to July. It is an important browse plant.

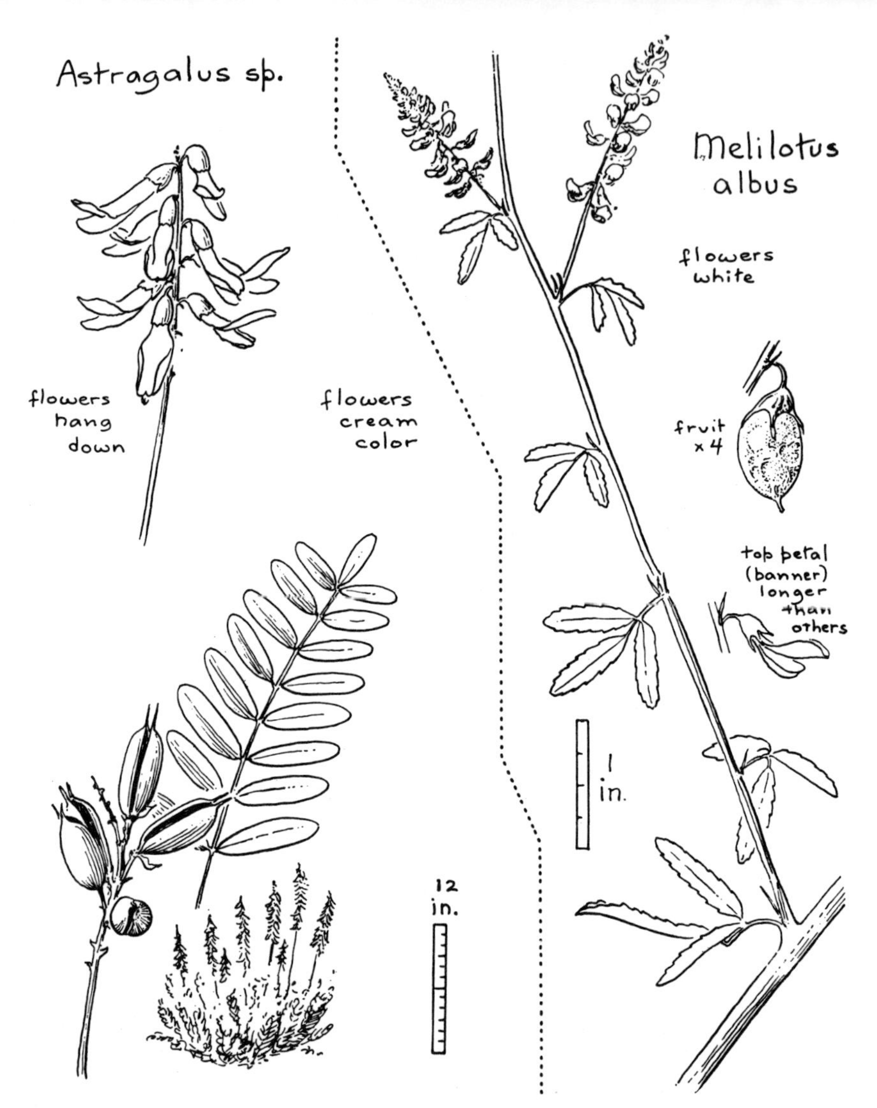

WHITE SWEETCLOVER (Melilotus albus)
Pea family *(Leguminosae)*. Blooms white, summer.

A tall, fragrant, roadside weed introduced from Europe and distributed throughout the United States. It often grows in solid masses for many miles, is tolerant of alkaline soil and is an excellent honey plant. It has been cultivated to some extent as a forage plant.

The yellow sweetclover is described on page 66.

LOCOWEED, MILKVETCH (Astragalus sp.)
Pea family *(Leguminosae)*. Blooms cream, May.

The milkvetches are a large group of plants and the species are difficult to identify. The drawing shows a species closely related to *Astragalus oophorus* (page 95), but with cream colored flowers and a pod that is not mottled. Common by the road between Los Alamos and Santa Fe, New Mexico.

Some species of the group cause loco poisoning in livestock, especially horses.

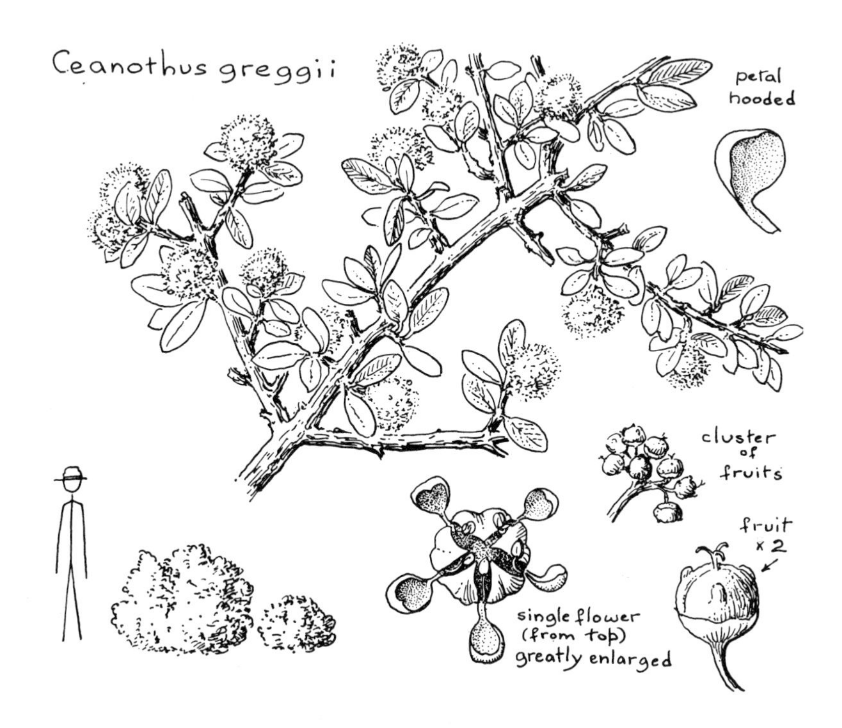

DESERT CEANOTHUS, WILD LILAC, BUCKBRUSH (Ceanothus greggii)

Buckthorn family *(Rhamnaceae)*. Blooms white, sometimes bluish or pinkish, March-April; sometimes on to September.

Western Texas to southern California and northern Mexico, 3,000-5,300 feet.

A woody shrub, with clusters of feathery white flowers, growing in thickets in the lower elevations of the Pinyon and Juniper Belt. It is not tall, seldom exceeding 5 feet. The branches are thick and rigid and grow at right angles.

This is the only local species of Ceanothus that has opposite leaves. The thick, rough leaves are grayish-green and small, about ½-¾ inch long. They are browsed by cattle and sheep, and especially by deer.

The small fragrant flowers grow in crowded clusters. The five petals are shaped like tiny hoods and have a narrow base or "claw." The long, thread-like stamens protrude beyond the petals, which gives a feathery appearance to the cluster. Nectar from the flowers makes a good honey.

The other species of buckbrush are common; *(Ceanothus martinii)* and *(Ceanothus fendleri)*. Both have alternate leaves. The former is not a spiny shrub and has thin leaves. It grows in Zion Canyon and at Grand Canyon.

The Fendler Ceanothus or New Jersey tea *(Ceanothus fendleri)* is small, spiny, and has thick leaf blades. It grows in Grand Canyon, Zion, and Mesa Verde National Parks and in Bandelier National Monument.

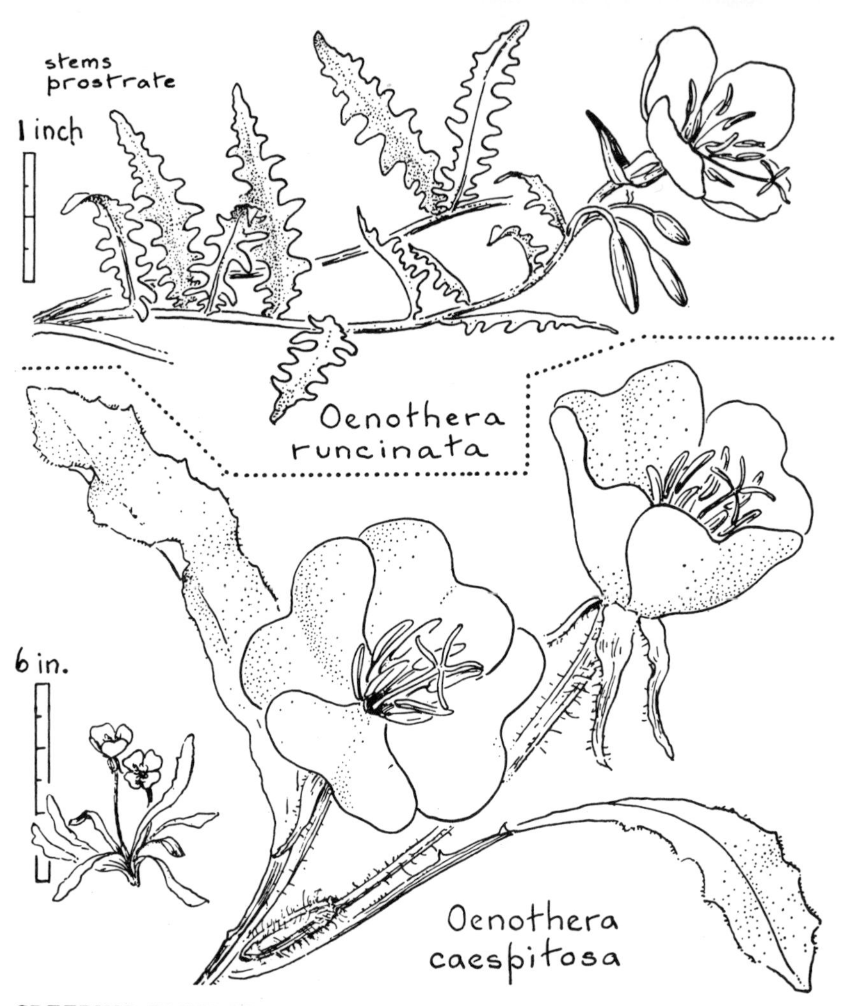

CREEPING PRIMROSE (Oenothera runcinata)

Evening-Primrose family *(Onagraceae)*. Blooms white, fading pink, May-June.

Arizona and New Mexico to Texas and Chihuahua, 4,500-7,500 feet.

This evening-primrose grows on dry plains and hills. Its leafy stems lie along the ground. There are many white flowers, smaller than those of the stemless primrose, and the plant is not as common. The drawing was made from a variety having stiff hairs. (var. *leucotricha*) collected near Santa Fe.

EVENING-PRIMROSE, STEMLESS PRIMROSE (Oenothera caespitosa)

Blooms white, summer.

Large, fragrant white flowers, common in the Southwest, which open in the evening, last only a few hours and turn pink as they wilt.

FIELD BINDWEED (Convolvulus arvensis) See Page 31.

Convolvulus family *(Convolvulaceae)*. Blooms white striped with pink; buds purplish pink, May-June.

Extensively naturalized from Europe, the plants are creeping and prostrate, with dull green leaves. A troublesome weed growing on roadsides and fields.

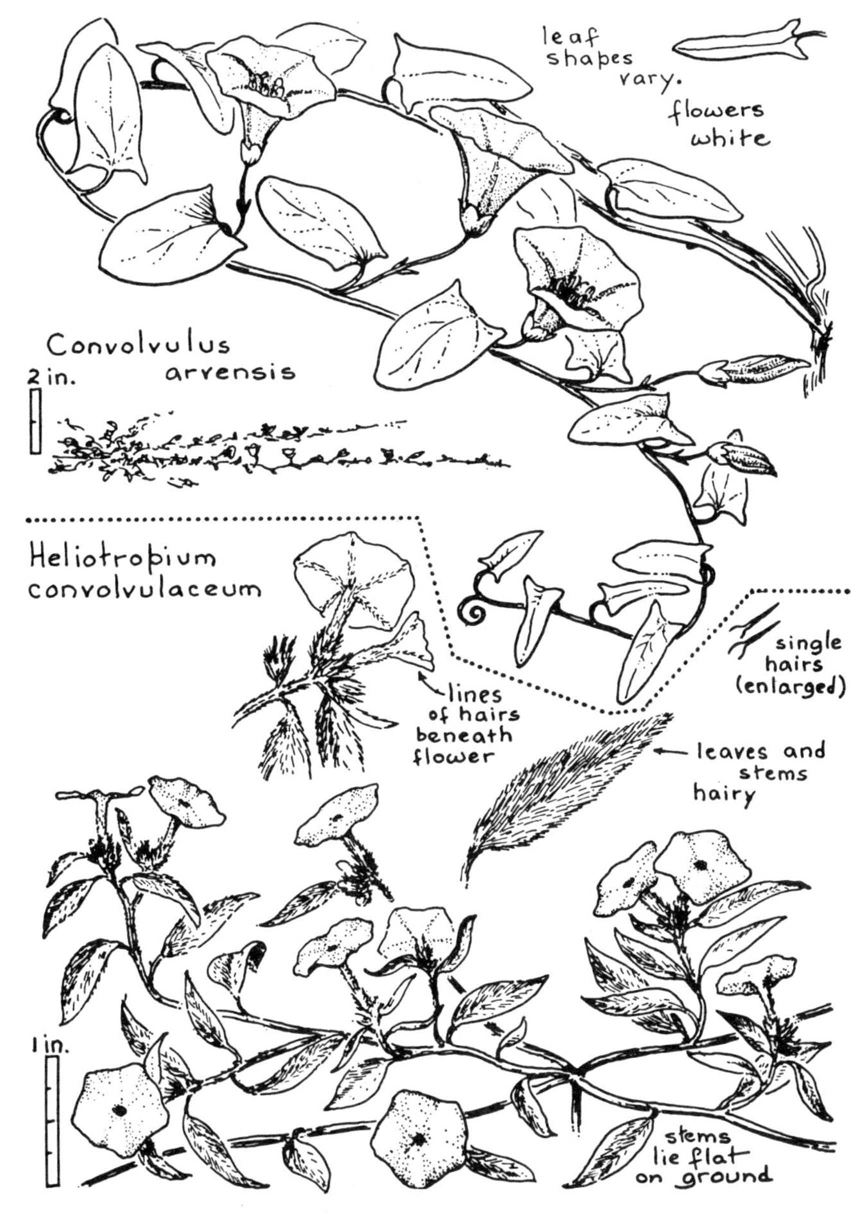

PHLOX HELIOTROPE (Heliotropium convolvulaceum)
Borage family *(Boraginaceae)*. Blooms white, mid-summer.
Nebraska to Texas, southern Utah, Arizona and Mexico, 4,500-6,000 feet.

Phlox heliotrope, a relative of the garden heliotropes, is a low spreading and freely branched herb that favors dry sandy places. It grows with the pinyons and junipers and also at lower elevations.

The sweet scented flowers, which open in the late afternoons, are pure white and there are many on a plant so that they make quite a showing along the roadsides. The fruit consists of four one-seeded nutlets.

This plant is well covered with rigid hairs that lie close to the stems, leaves, and even the flower tube.

The phlox heliotrope is very common, growing through northern New Mexico and northeastern Arizona.

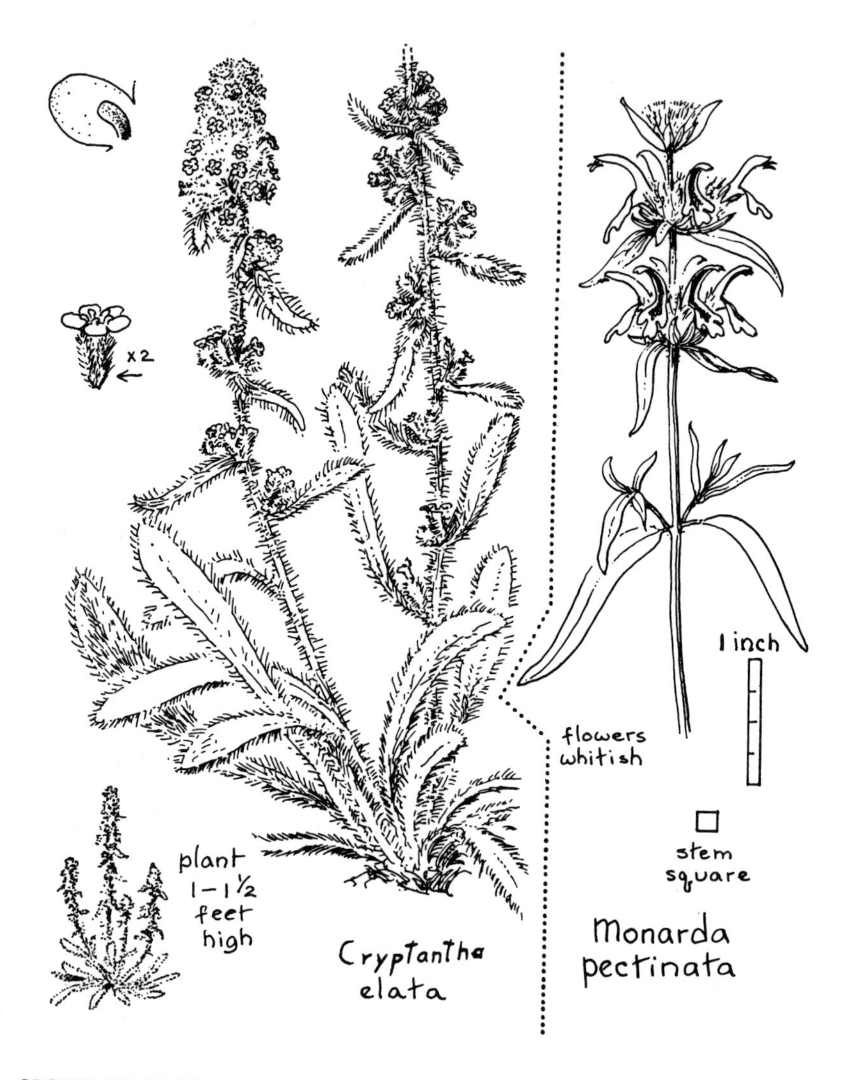

CLIFFDWELLERS CANDLESTICK (Cryptantha elata)
Borage family *(Boraginaceae)*. Blooms cream, June. 5,000-6,000 feet.

May be seen with sagebrush on clay hills. It is common at Mesa Verde National Park where it was given the name of cliffdwellers candlestick.

PONY BEE BALM, WHITE HORSEMINT (Monarda pectinata)
Mint family *(Labiatae)*. Blooms white, late summer.
Nebraska and Colorado to Texas and Arizona, 5,000-7,000 feet.

The white horsemint grows in sandy soil. It is found on the south rim of Grand Canyon and at Bandelier National Monument. The flowers grow in groups along the stem at the base of the leaf.

A purple flowered horsemint, or bergamot, is described on page 101.

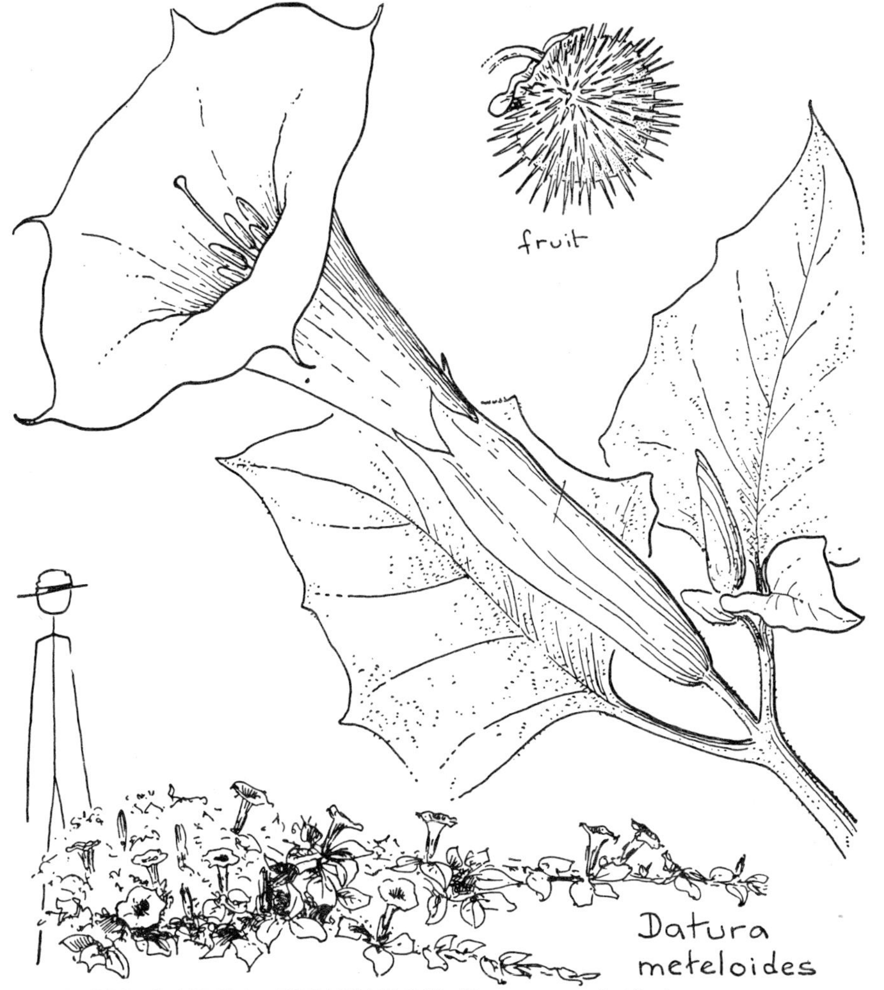

SACRED DATURA, THORNAPPLE (Datura meteloides)
Potato family *(Solanaceae)*. Blooms white tinged with lavender, May-October. Colorado to Texas, Arizona, southern California and Mexico, 1,000-7,000 feet.

This is a conspicuous, coarse herb with many beautiful, large, funnel-shaped white flowers, growing on dry slopes and in arroyos of the Pinyon-Juniper Belt, and at lower elevations with the creosotebush.

Datura is a large, gray-green herb, and forms a spreading clump not over 4 feet high. The simple alternate leaves are velvety beneath and have an unpleasant odor.

There are many fragrant, five-parted flowers, sometimes 100 on a plant, which open in the early evening and close in the forenoon of the next day. It is sometimes called moon lily, although it is not related to the lilies.

The fruit is a round, nodding pod thickly armed with short equal prickles. Seeds are light brown.

Datura is poisonous, containing the drug atropine. California Indians sometimes use it for medicinal purposes and as a narcotic to induce dreams and visions in religious ceremonies.

Occurs in Grand Canyon, Zion, and Petrified Forest National Parks, and in Bandelier, Chaco Canyon, Pipe Springs, Canyon de Chelly, Aztec Ruins, Wupatki, and Montezuma Castle National Monuments.

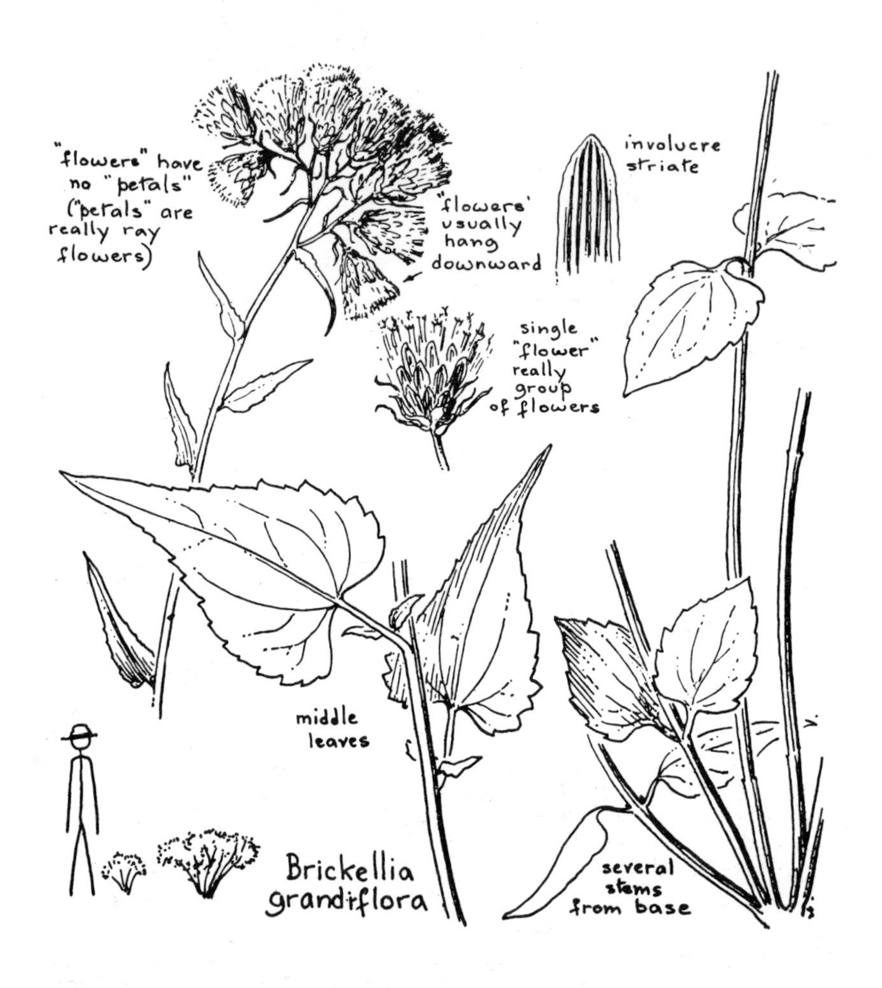

TASSELL-FLOWERED BRICKELLIA, BRICKELLBUSH, SHEATH FLOWER (Brickellia grandiflora)

Sunflower family *(Compositae)*. Blooms white, August-October.

Missouri and Arkansas to Montana and Washington, south to New Mexico, southern Arizona and California, 5,000-9,000 feet.

This is a somewhat shrubby plant with green stems and thin, arrowhead-shaped leaves. The tiny white flowers are crowded together in a head; none of them have petal-like rays. It is more attractive after it has gone to seed, when the down on the seeds is white, and the drooping flower head looks like a tassel.

Brickellia grows in rich soil by streams, often under the pines. At Grand Canyon it is seen near Cliff Springs on the north rim and in Long Jim Canyon on the south rim. In Zion Canyon it occurs along the Narrows Trail, in Mesa Verde Park in Navajo Canyon, and in Bandelier National Monument near the head of Alamo Canyon.

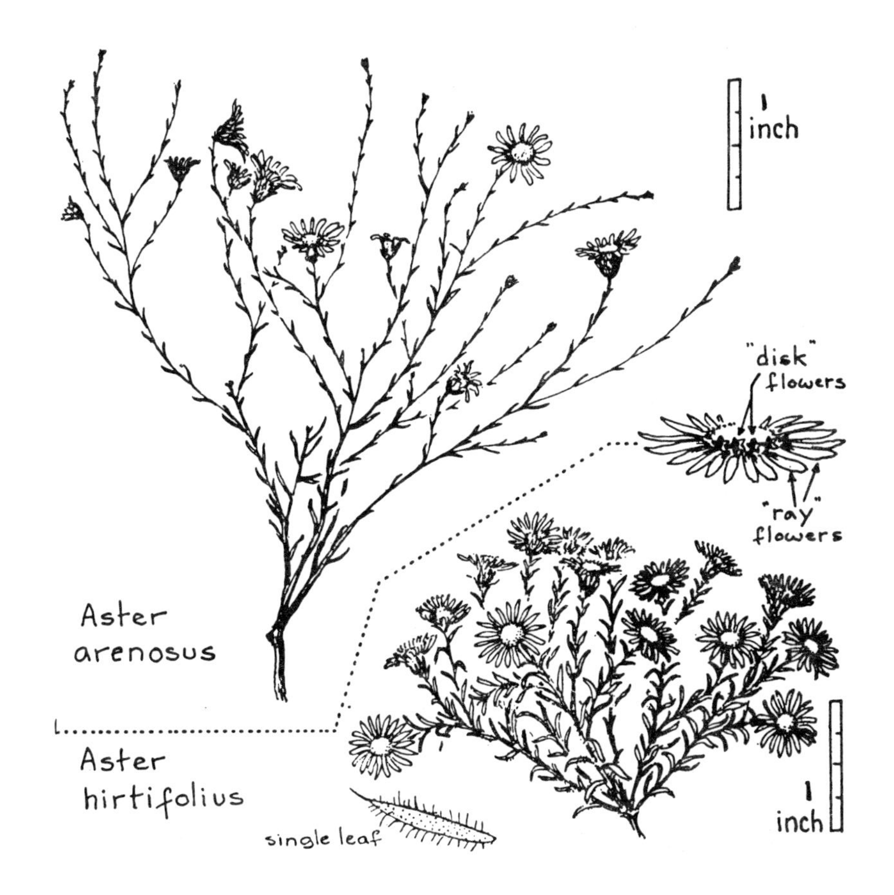

HEATH-LEAVED ASTER, BABYWHITE ASTER (Aster arenosus)
Sunflower family *(Compositae)*. Blooms white, fading pink or purplish, April-October.
Colorado to Texas, Arizona and Mexico, 4,000-6,500 feet.

Leaves very small, not glandular; the upper ones scale-like. Much branched from a woody base.

Occurs in Grand Canyon, Mesa Verde, and Petrified Forest National Parks, and at Aztec Ruins and Chiricahua National Monuments. Also occurs at Santa Fe, New Mexico.

BABY ASTER (Aster hirtifolius)
Blooms white, fading pink or purplish, April-August.
Wyoming to Nevada, south to Texas and Arizona, 5,000-7,000 feet.

Leaves glandular, edges of leaves fringed with tiny hairs. Plant low growing, in many-flowered tufts.

Common at Grand Canyon in the village and along the trails. Also grows at Navajo, Chaco Canyon and Aztec Ruins National Monuments. This species is probably the same or close to *Aster leucelene,* listed for Zion Canyon, Canyon de Chelly, and Wupatki.

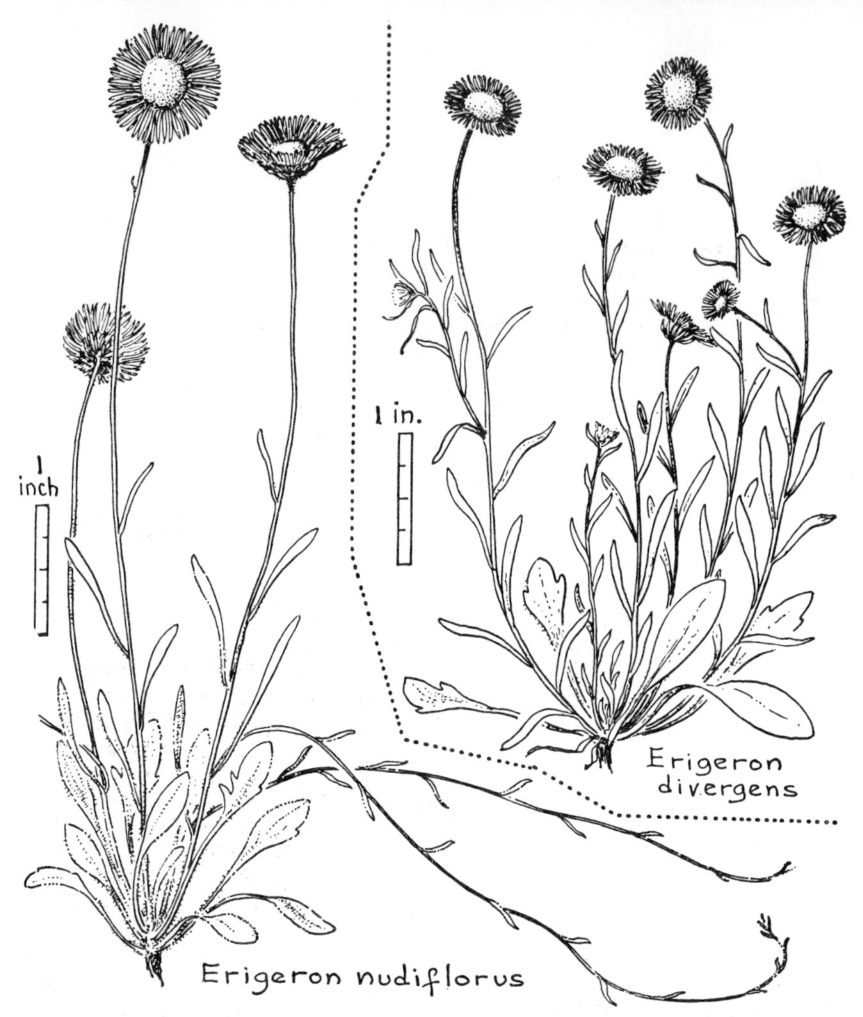

SPREADING DAISY, BRANCHING FLEABANE (Erigeron divergens)
Sunflower family *(Compositae)*. Blooms white, violet or purplish, February-October.
South Dakota to British Columbia, south to Texas, New Mexico, Arizona, California and northern Mexico, 1,000-9,000 feet.

Common and pretty, this daisy has a very wide range. It grows on dry rocky slopes and mesas and in open pine woods. It is found in nearly all of the National Parks of this region.

The flower head has a compact, bright yellow center. The many petal-like rays are narrow—white towards the center—and violet, bluish or purplish towards the ends.

SPRAWLING DAISY (Erigeron nudiflorus)
Blooms white or pink, March-July.
Colorado and Utah to Texas, New Mexico, Arizona and northern Mexico, 4,000-7,000 feet.

This species is similar to the one above but has larger, white flower heads. The young flowering stems do not have leaves and the plant puts out long, spreading runner-like branches, near the ground, which are leafy but do not bear flowers.

This plant is much like *Erigeron flagellaris* but has a larger flower head.

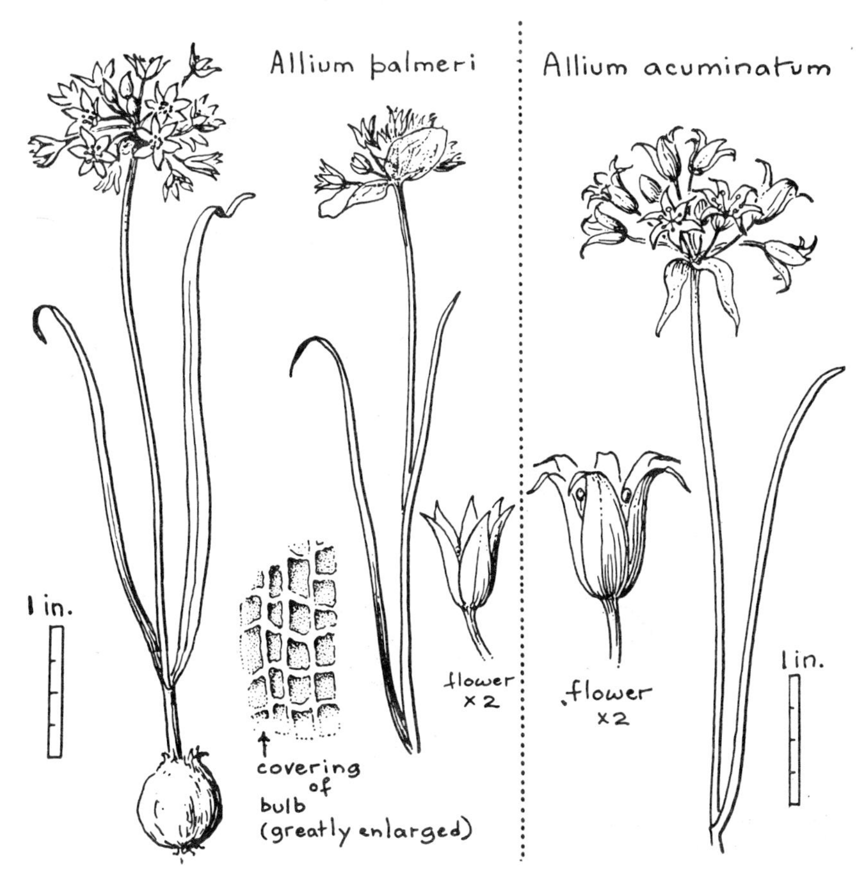

WILD ONION (Allium palmerii)
Lily family *(Liliaceae)*. Blooms light rose, May-July.
Southern Utah, New Mexico and Arizona, 4,000-7,500 feet.

This wild onion has delicate, starry, rose-pink flowers which are gathered into an umbrella-shaped cluster at the top of a leafless stalk, attaining a height of about 6-12 inches. It may be seen at Grand Canyon in shady places among the rocks.

The thin, papery bracts which envelope the flower cluster before blooming persist at the base of the cluster after the flowers open. The six separate petals are delicately striped and flower stalks are usually two to a plant. The two narrow, thickish leaves grow at the base of the flower stalks.

The bulb, with an onion smell and taste, is coated in layers. The outer coat is light brown and has small ridges.

Wild onion is related to the domestic onions, garlic, leek, and chives. Indians formerly used bulbs for food and seasoning, raw or after heating in ashes. They were sometimes stored for winter use.

An even more beautiful wild onion *(Allium acuminatum)*, occurring at slightly higher elevations, blooms earlier in the spring. It has a larger flower, deep rose-pink in color.

This is the common wild onion at Mesa Verde and Zion National Park. The nodding onion *(Allium cernuum)* occurs at Grand Canyon National Park and at Bandelier National Monument. Flowers are pale pink or nearly white. Another species *(Allium macropetalum)*, common in Petrified Forest National Park and in Navajo and Aztec Ruins National Monuments, has orchid-pink flowers, and is a low growing herb.

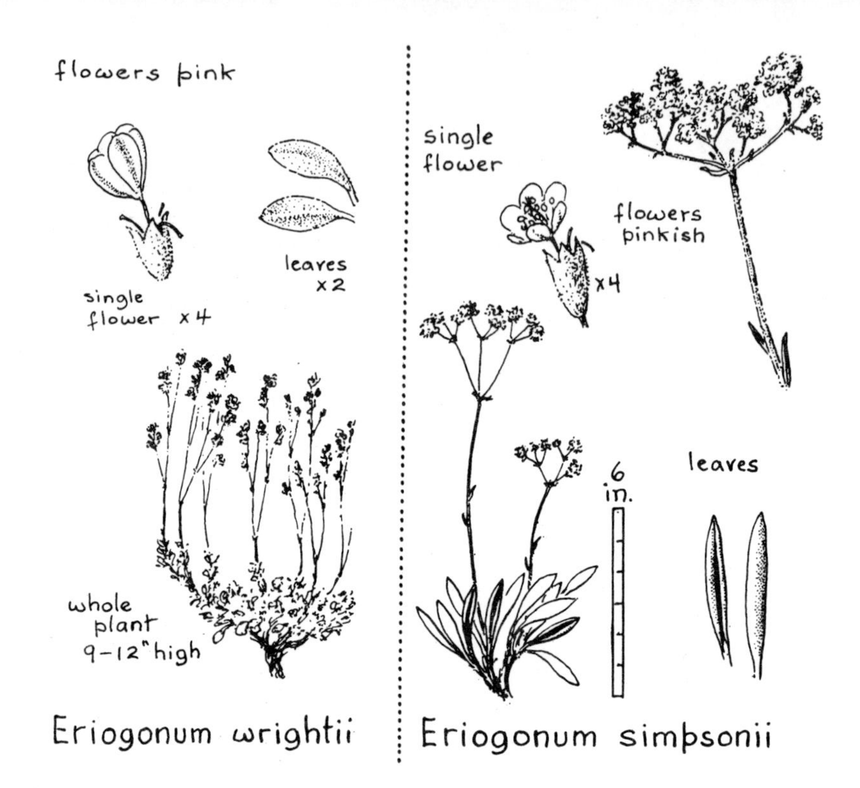

WILD BUCKWHEAT (Eriogonum)
Buckwheat family *(Polygonaceae).*

Buckwheats form one of the largest groups of plants in the Southwest. Three common species are mentioned here, two with yellow flowers on page 57. They all have many colorful flowers in umbrella or spike-shaped arrangements; their larger leaves are usually in a rosette at the base of the plant and are often covered with white down. Buckwheat and rhubarb are two very useful members of this family.

WRIGHT BUCKWHEAT (Eriogonum wrightii)
Blooms pink or white, June-October.
Colorado to Texas, Arizona, California and northern Mexico, 3,000-7,200 feet.

A very common perennial which decorates the roadsides for many miles through the Southwest with its delicate pink flowers. The flowers appear in clusters on spikes at the end of the branches. Leaves are scattered along the woody stems.

It is a good browse plant for cattle and deer and also yields a fine, almost colorless honey.

PINK BUCKWHEAT (Eriogonum simpsonii)
Blooms white ribbed with red, July-September
Colorado to Nevada, New Mexico and Arizona, 3,500-7,500 feet.

Pink buckwheat has some leaves on the branches. Leaves are greenish above and white woolly below.

It is found on both rims of Grand Canyon, at Petrified Forest National Park, and at Navajo National Monument.

SAND-VERBENA (Abronia nana)
Four-o'clock family *(Nyctaginaceae)*. Blooms pink or whitish, April-May. Southern Utah and Nevada and northern Arizona, 3,000-5,000 feet.

This pretty pink, fragrant flower is found in Grand Canyon, blooming in the early spring. It occurs, but not commonly, in the lower elevations of the Sagebrush-Pinyon Belt. The sand-verbena is not to be confused with the true verbena which is in the vervain family and not at all related to the four-o'clock family.

The five-lobed, tubular flowers are clustered together in heads at the base of which are thin, papery, whitish bracts. The central parts of the flower are hidden within the tube; the long stalk supporting the flower heads arises from the stem in the angle of the leaves.

The sand-verbena is a low growing plant with a thickened, woody root. The toothless leaves are opposite and thick. The fruit is leathery and has narrow, thickish, opaque wings.

Other species of sand-verbena are reported from National Parks and Monuments as follows:

Abronia pumila .... **Navajo**
Abronia fragrans .... **Navajo, Aztec Ruins, El Morro, Petrified Forest, Mesa Verde**
Abronia elliptica .... **Navajo, Aztec Ruins, Chaco Canyon, Wupatki, Zion**
Abronia salsa .... **Zion**
Abronia cycloptera .... **El Morro**

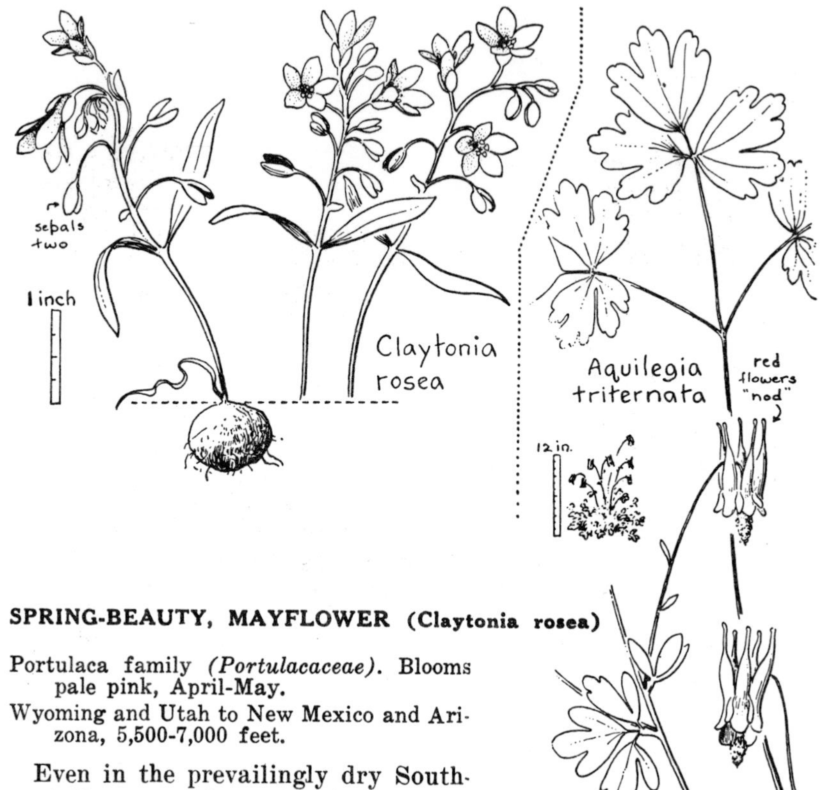

SPRING-BEAUTY, MAYFLOWER (Claytonia rosea)

Portulaca family *(Portulacaceae)*. Blooms pale pink, April-May.

Wyoming and Utah to New Mexico and Arizona, 5,500-7,000 feet.

Even in the prevailingly dry Southwest there are places for little spring flowers that love moist rich soil and shade. This fragile spring-beauty frequents the Narrows, Hidden, and Refrigerator Canyons in Zion National Park. It is found on Buggeln Hill on the south rim of Grand Canyon and at Aztec Ruins National Monument.

A similar species *(Claytonia lanceolata)* with notched petals grows in Mesa Verde National Park where Spruce and Spruce Tree Canyons join.

The spring-beauty is a small, succulent plant about 3-4 inches high, with a reddish stem emerging from a solid, round bulb. The delicate flowers open only in the sunlight and last but one day.

The pale, pink petals have darker pink or purple veins. At the base of each petal is fastened a pink stamen. The pistil is also pink; the style is three-cleft. The flower rests in two yellowish-green sepals that persist on the plant after the petals fall.

RED COLUMBINE (Aquilegia triternata)

Buttercup family *(Ranunculaceae)*. Blooms red, sometimes partly yellow, June-October.

Colorado, New Mexico and Arizona, 6,000-10,000 feet.

A species much like the red-and-yellow columbine of higher elevations, *Aquilegia elegantula,* has been observed at Ribbon Falls and in Long Jim Canyon in Grand Canyon National Park.

The yellow columbine is shown on page 59.

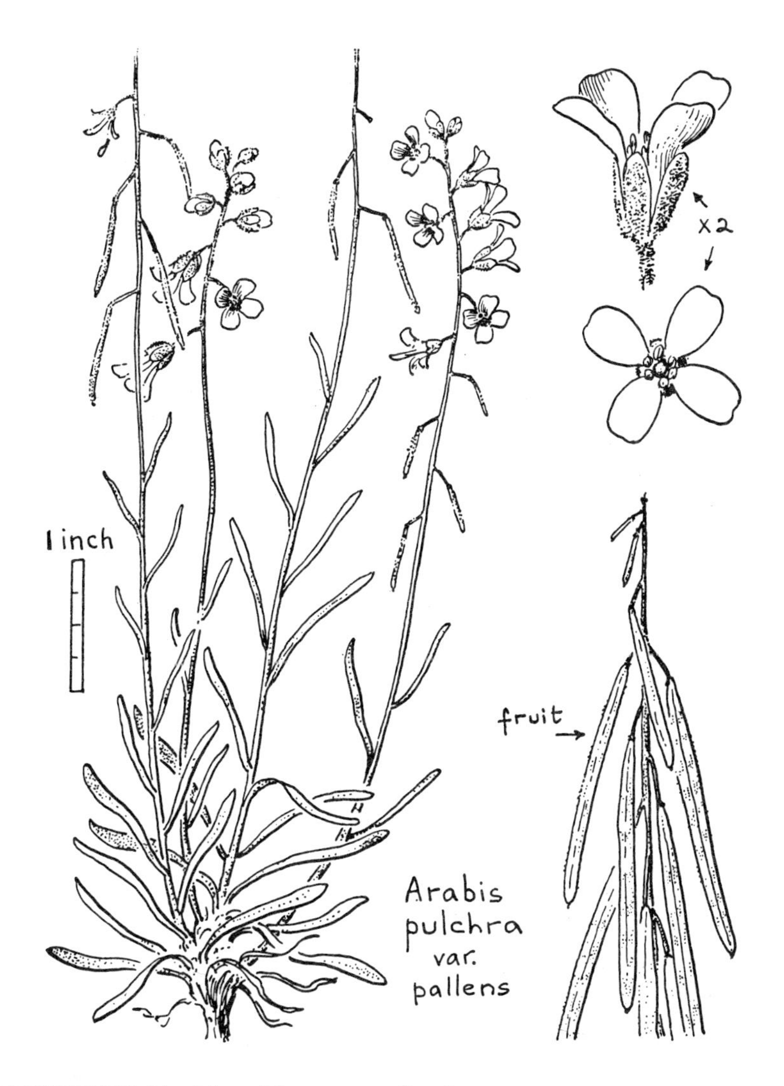

ROCKCRESS (Arabis pulchra var. pallens)
Mustard family *(Cruciferae)*. Blooms pink to purple, May-July.
Western Colorado, eastern Utah and northeastern Arizona, about 6,000 feet.

Rockcress frequents sandy hills and open knolls with sagebrush, pinyon, and juniper. Its purplish-pink flowers are relatively large and showy and are arranged in a spike. They have four petals and six stamens. The name of the mustard family *(Cruciferae)* comes from the cross-like arrangement of the four-petaled flowers.

The seed pods are long and flat, and are not crowded on the flower stalk. The seeds are broadly winged.

The leaves are bluish in color; those on the stem are rather numerous, and those at the base of the plant form a rosette. The leaves, pods and calyx are covered with fine, star-shaped hairs.

Rockcress grows at Grand Canyon and Mesa Verde National Parks, and at Navajo and Colorado National Monuments.

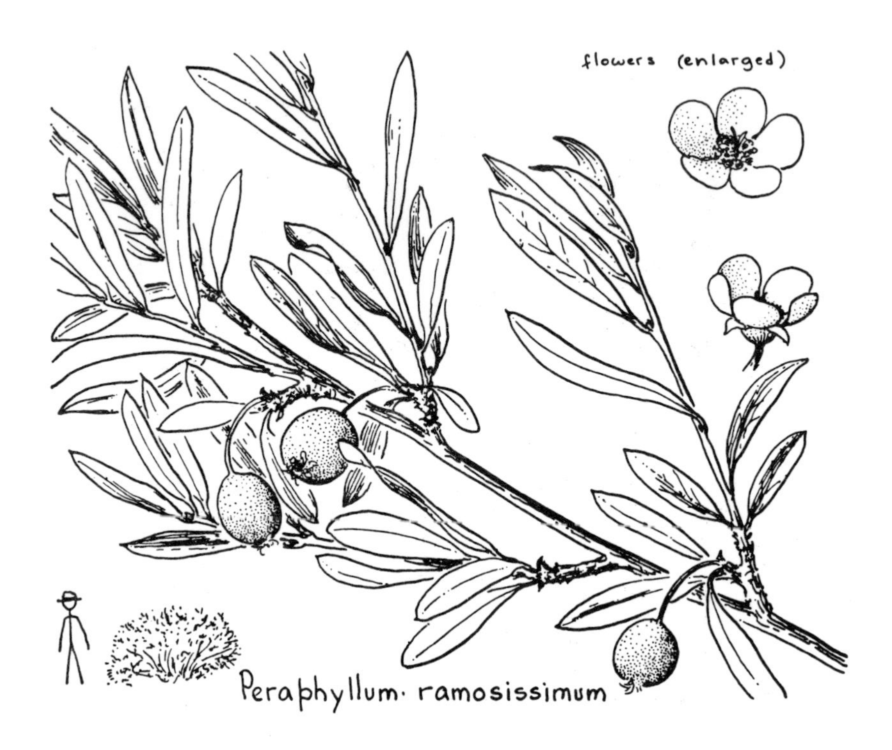

SQUAW APPLE, WILD CRAB APPLE (Peraphyllum ramosissimum)
Rose family *(Rosaceae)*. Blooms pink, May-June.
Oregon and California to western Colorado and New Mexico, 5,500-8,000 feet.

This is a rather rare but very pretty shrub with its pink, apple-blossom-like flowers. It is intricately branched, with grayish bark, and grows from about 3 to 6 feet high. It inhabits dry hills of the Pinyon-Juniper Belt and extends into the far Northwest.

The squaw apple is closely related to the serviceberry *(Amelanchier)*. Its comparatively large, pink blossoms are made up of many stamens and two styles. The fruit looks like a small apple, and is yellow, tinged with purple. It has very bitter juice.

The flowers appear on the bush at the same time as the leaves, which have silky hairs. They grow singly or sometimes in small bunches. Branches are short and rigid.

In Zion National Park the squaw apple is found only on the east rim. It is rather conspicuous in Mesa Verde National Park, growing northwest of the Park Service headquarters and near the Park entrance. It may also be seen south of Durango, Colorado, near the New Mexico border, and it is reported from the vicinity of Farmington, New Mexico.

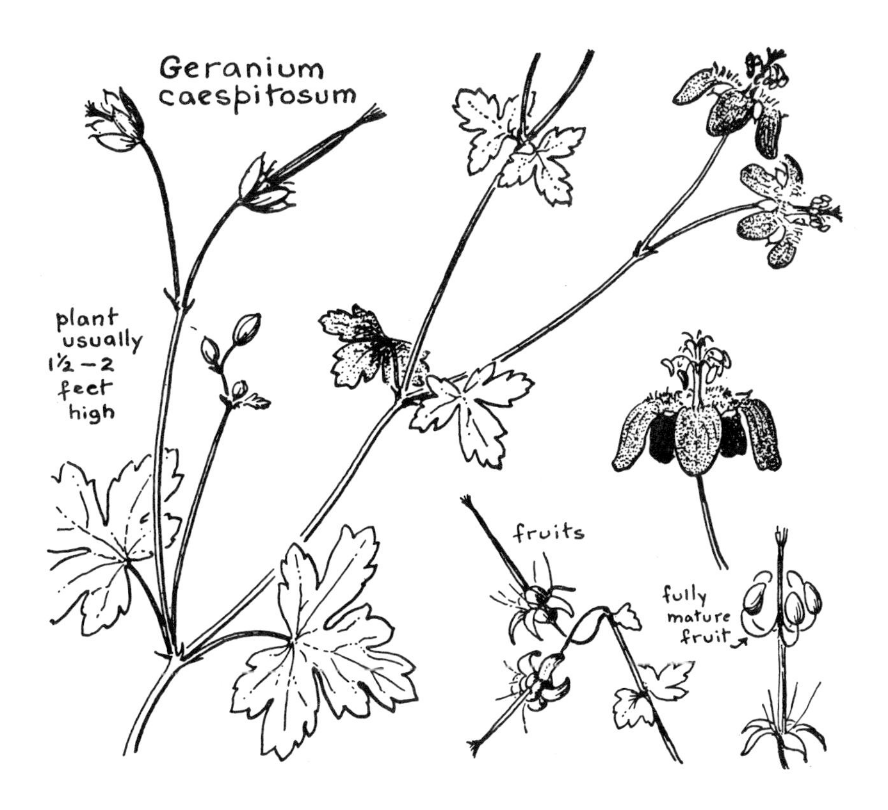

WILD GERANIUM, CRANESBILL (Geranium caespitosum)
Geranium family *(Geraniaceae)*. Pink to rose-purple, June-September. Colorado, New Mexico, Arizona, and Mexico, 5,000-8,000 feet.

This wild geranium, a common flower with a wide climatic range, prefers rich soil with the sagebrush and pinyon and also with the ponderosa pine.

The slender, sparsely branched plant emerges from a stout, woody taproot. The deeply toothed leaves are opposite and the leaf stems are long.

Flowers are a deep purplish-pink, the veined petals curve downward and are considerably longer than the sepals, and there are 10 little stamens.

The arrangement of the seed pods of the geraniums is unique: they are attached to the base of the long, persistent styles which surround a central beak-like column. Upon maturity these styles separate and curl upward, bringing the pods with them, at the same time scattering the seeds.

This species of geranium may be observed in the Santa Fe area, at Bandelier, Aztec Ruins, Navajo and Walnut Canyon National Monuments and at Mesa Verde and Grand Canyon National Parks.

Wild geraniums provide good forage for sheep. The rootstock of an eastern species *(Geranium maculatum)* is used medicinally as an astringent. Our cultivated geraniums *(Pelargonium)* came originally from South Africa.

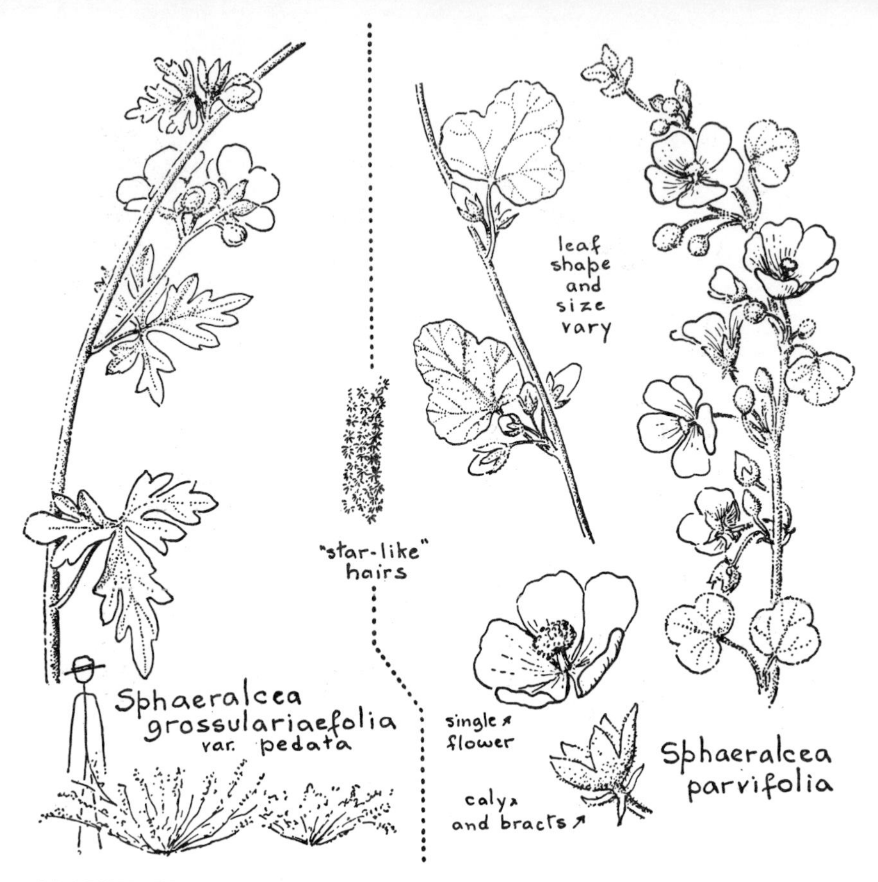

GLOBEMALLOW, SORE-EYE-POPPY (Sphaeralcea)

Mallow family *(Malvaceae)*. Blooms coral-red or pink, May-October, in the spring and later after summer rains.

Globemallows grow abundantly all through the Southwest. The many-flowered, curved stems sway gracefully in the wind, displaying the coral-red flowers which give vivid coloring to the landscape.

It belongs to the same family as the cultivated cotton, okra, and the ornamental hibiscus and hollyhock. The European marshmallow used in making the well known confection also is a member of this group.

The Hopi prize the globemallow medicinally and the sticky stems are used by several Indian tribes as a substitute for chewing gum. Sheep and goats browse the plants.

Two common species are illustrated here, easily distinguished by the shape of the leaf.

SMALL-LEAF GLOBEMALLOW (Sphaeralcea parvifolia)

Western Colorado to New Mexico, Arizona and eastern California, 4,000-7,000 feet.

Stems and leaves whitish or grayish. Shallowly lobed leaf, broad rounded lobes. Flowers coral-red to pink.

GOOSEBERRY GLOBEMALLOW (Sphaeralcea grossulariaefolia var. pedata)

Idaho and Washington to New Mexico, Arizona and California, 3,000-6,000 feet.

Flowers brick red to scarlet. Leaf deeply divided.

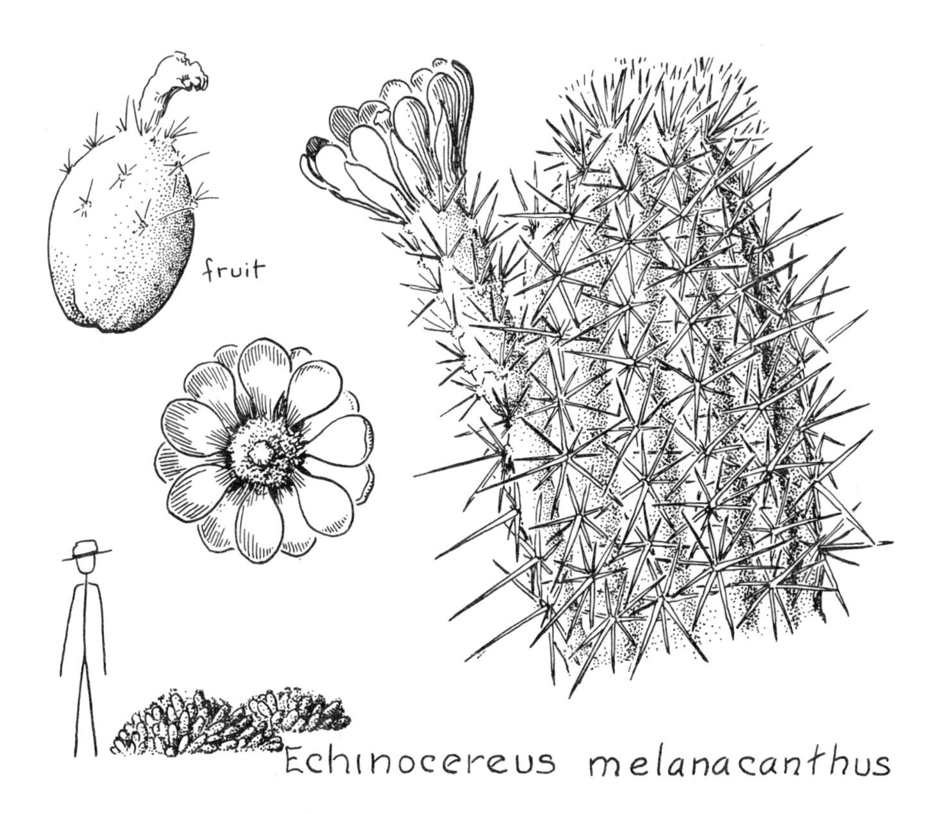

WHITE SPINED CLARET CUP, HEART TWISTER (Echinocereus melanacanthus) Echinocereus triglochidiatus var. melanacanthus.

Cactus family *(Cactaceae)*. Blooms crimson, May-July.

Colorado, Utah, New Mexico and Arizona, 4,000-9,000 feet.

This hedgehog cactus forms large mounds, sometimes as many as 50 stems in a mound, making a very effective display when it is in bloom—a mass of bright red flowers.

A common plant of the Pinyon-Juniper Belt, it may be seen in Mesa Verde and Zion National Parks and near Aztec Ruins National Monument. It grows on the lower elevations of the Kaibab Plateau.

The ribbed, cylindrical stems are yellow-green and about 4-6 inches long. The yellow-spines are straight and short. The edible and succulent fruit is yellow-green, about 1 inch long, has a thin rind, and the spines drop off as ripening occurs.

The flowers come out along the side of the stem and do not close at night as do the blossoms of some species of cacti.

The heart twister is the name the Navajo give to this cactus. According to legend, in their wanderings they were made ill by eating the juicy fruit of this hedgehog cactus which has a strawberry-like odor. It is, therefore, customary for them to offer a hair of the head to the "Cactus People" before they eat the fruit to prevent their hearts from being twisted with pain.

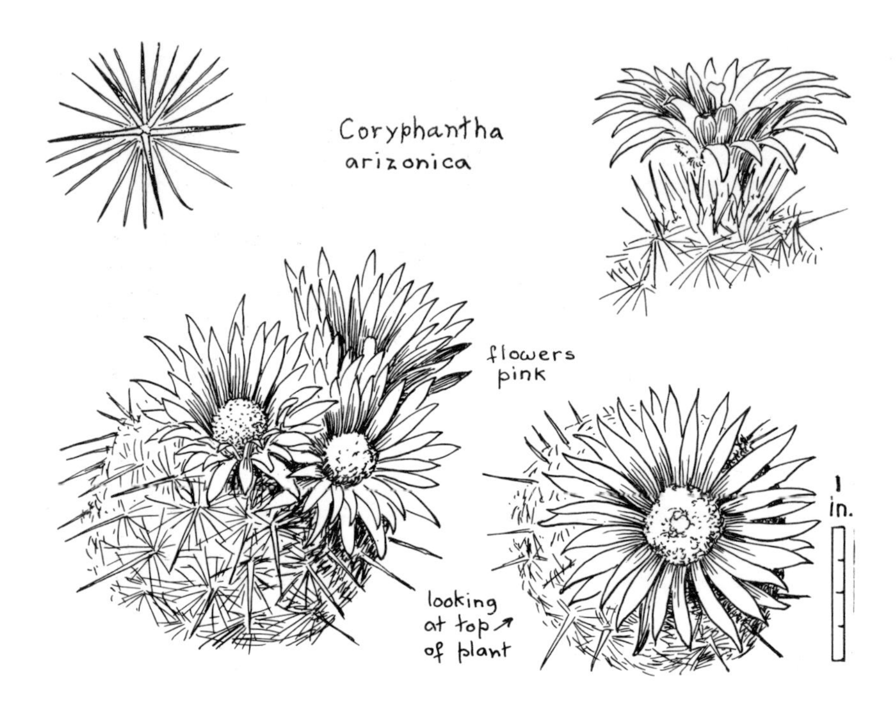

BEEHIVE CACTUS, ARIZONA PINCUSHION CACTUS (Coryphantha arizonica) Coryphantha vivipara var. arizonica.
Cactus family *(Cactaceae)*. Blooms deep pink, May-June.
Southwestern Utah to Northern Arizona, 4,000-8,000 feet.

The spines of this cactus are brown and white and grow on the tips of projecting "tubercules," not continuous ribs like those of the species described on page 45. The stems are solitary at first, later branching to form large mounds.

Beautiful pink flowers come out at the top of the plant. The fruit is green, sometimes tinged with red.

This pincushion cactus prefers rocky places in the pinyon-juniper woodland and in the ponderosa forests. It is found in Grand Canyon, Zion, and Petrified Forest National Parks.

POINTLEAF MANZANITA, BEARBERRY (Arctostaphylos pungens)
Heather family *(Ericaceae)*. Blooms white or pink, March-April.
New Mexico, southern Utah, Arizona, southern California and Mexico, 4,000-8,000 feet.

The manzanita, a relative of the rhododendrons, has crooked, spreading branches and often grows in a thicket or "chaparral." It is also closely related to the creeping evergreen plant, "kinnikinnick" *(Arctostaphylos uva-ursi)* which grows on the forest floor at higher elevations.

This species is common at Grand Canyon. Its smooth, mahogany-red bark, pale or bluish-green upturned leaves, and clusters of nodding, pink flowers make it a decorative shrub.

The fruit is berry-like; the soft pulp surrounds several nutlets. Bears, chipmunks, and other animals eat the berries, and jelly can be made from them before they ripen. Manzanita is a Spanish word meaning "little apple," which is a good description of the berry-like fruit.

The leathery painted leaves are evergreen; and by twisting their stalks they assume a vertical position, thus avoiding unnecessary evaporation.

A similar plant is the "greenleaf manzanita" *(Arctostaphylos patula)*, which also grows at Grand Canyon and in Zion Canyon. It has larger, brighter green, oval leaves which are rounded at the tip. The flowers grow in a more open cluster.

Deer, mountain sheep, and goats browse on the leaves. Indians use the berries for food and for making a pleasant, acid drink.

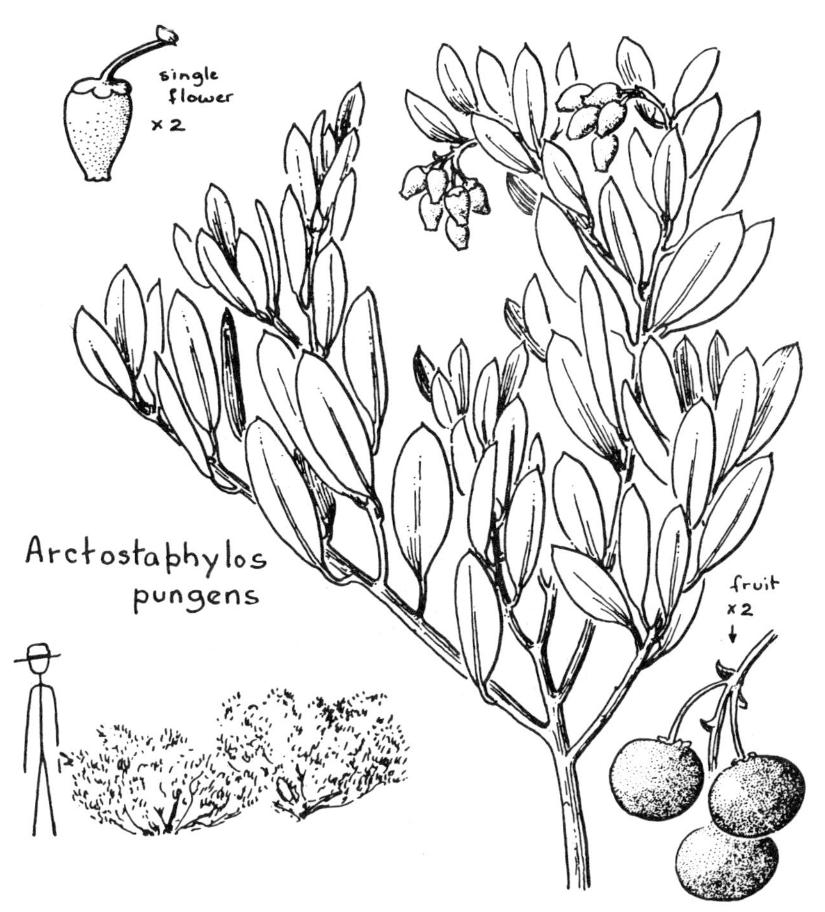

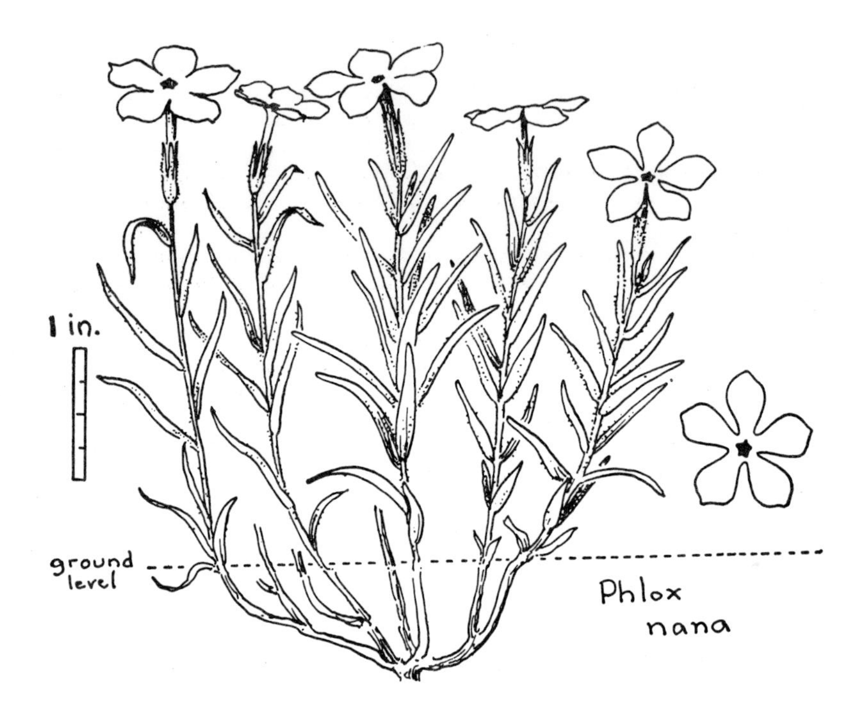

PINK PHLOX (Phlox nana)

Phlox family *(Polemoniaceae)*. Blooms deep rose-pink, spring and summer. Western Texas, New Mexico, southeastern Arizona and Chihuahua, 5,500-7,500 feet.

The phlox is a familiar flower since it is a great favorite and the wild phlox looks very much like the cultivated forms. This particular one grows from a perennial tap root and branches mostly underground so that the unbranched stems look like individual plants. Sheep seem to relish the flowers.

Rose-pink flowers grow singly at the tops of the branches. The five wide petal lobes are rolled up in the bud, but when the flower opens they spread out flat. Stamens are hidden within the flower tube, but if one were to cut the tube open he would see the five stamens of different lengths fastened to the walls of the tube between the lobes. The seeds of phlox are not sticky, as are those of the somewhat similar gilia plants.

The stems, leaves and calyx are slightly hairy with tiny glands on the ends of the hairs.

This is a common species in the pinyon-juniper woodland. It grows on the hills around Santa Fe and blooms in the spring and again after the summer rains.

A similar species of phlox *(Phlox longifolia)* grows in Grand Canyon, Mesa Verde and Zion National Parks and in Montezuma Castle and Bandelier National Monuments.

It is a taller plant than the species described above, the petal lobes are narrower, the leaves are longer and more narrow, rather stiff, and grey-green in color. The color of the flower ranges from deep pink to white. Two stamens are visible above the flower tube.

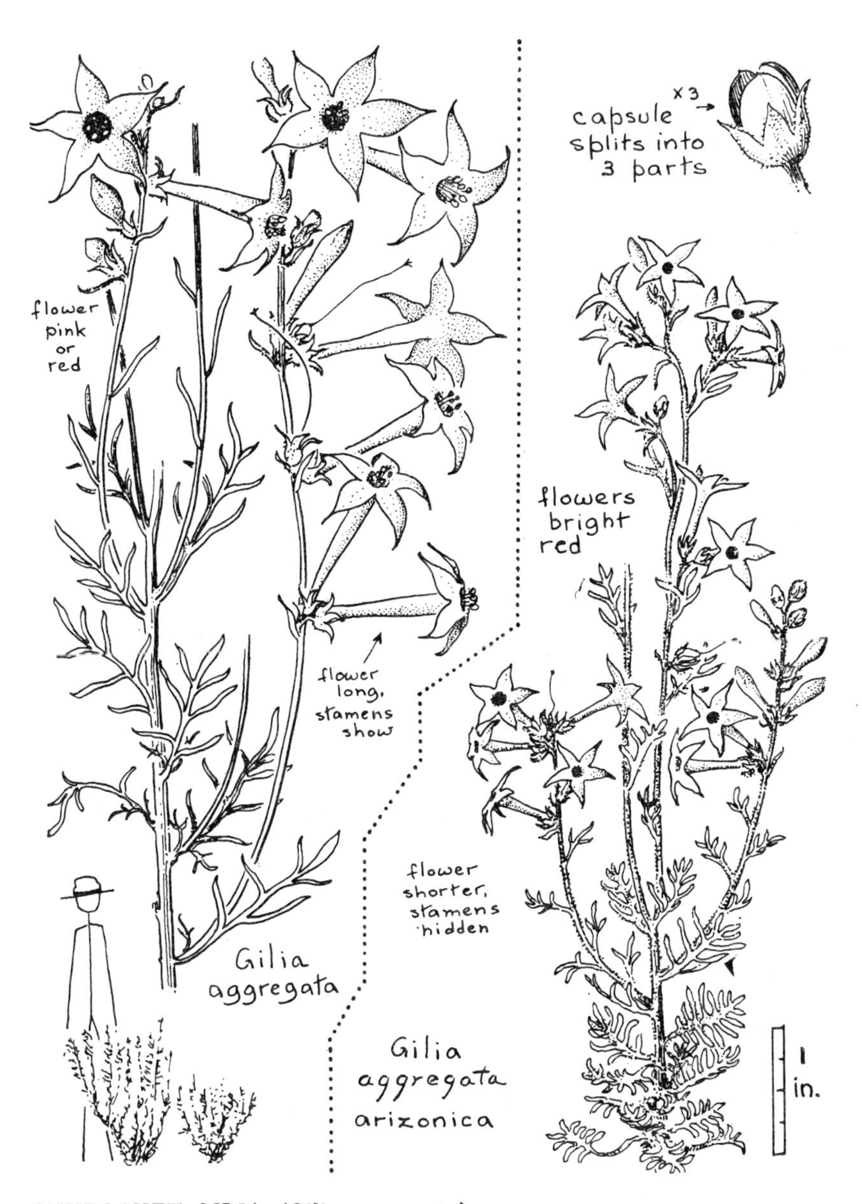

SKYROCKET GILIA (Gilia aggregata)
Phlox family *(Polemoniaceae)*. Blooms deep pink sometimes nearly red, with white dots, June-August.
Montana to British Columbia, south to New Mexico, Arizona and California, 5,000-8,000 feet.

Very showy. Attracts hummingbirds.

RED GILIA (Gilia aggregata var. arizonica)
Blooms bright red with white dots, May-September.
Southern Utah and Nevada and northern Arizona, 5,000-7,500 feet.

A blue flowered gilia is shown on page 99.

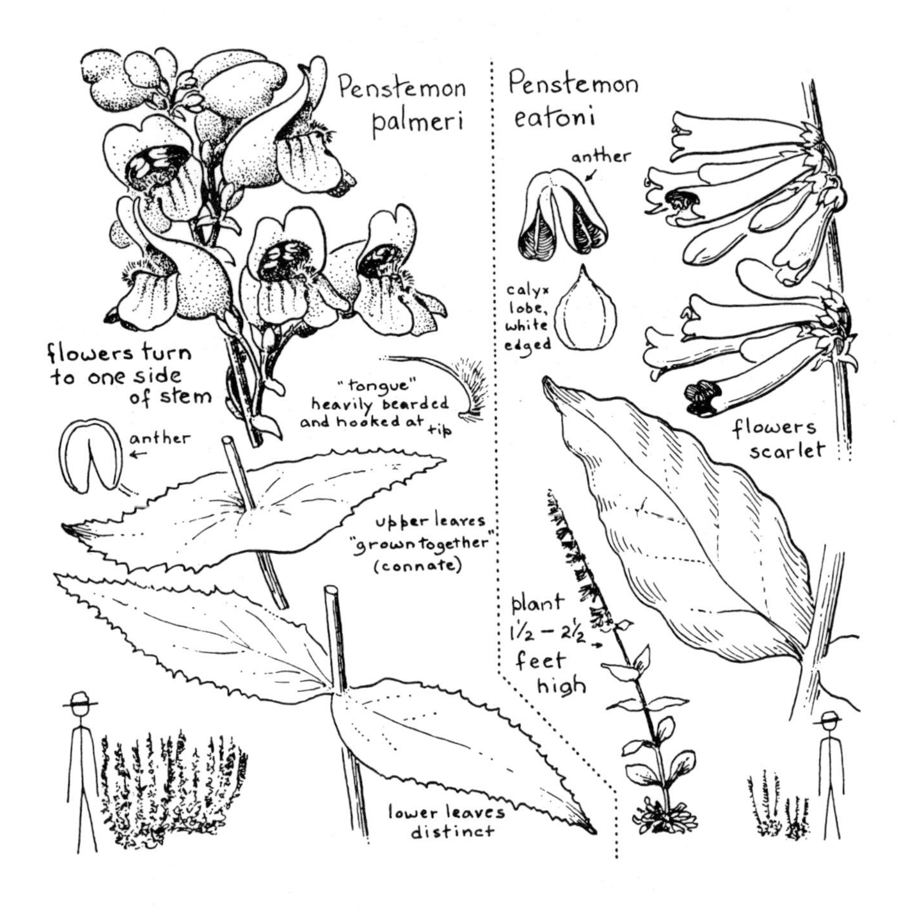

PALMER PENSTEMON, PINK WILD SNAPDRAGON (Penstemon palmeri)
Snapdragon family *(Scrophulariaceae)*. Blooms white tinged with pink, or pale lavender, May-June.
Utah and Arizona to California, 3,500-6,500 feet.

The penstemons or wild snapdragons are among the most beautiful and showy flowers of the Southwest mesas. The Palmer penstemon is one of the few fragrant ones; it is the largest and most common wild snapdragon growing with sagebrush and pinyon, especially spectacular on the floor of Zion Canyon.

SCARLET BUGLER, RED PENSTEMON (Penstemon eatoni)
Blooms scarlet, March-June.
Southwestern Colorado to central Arizona and California, 2,000-7,000 feet.

The scarlet bugler is brilliant on mesas, along roadsides and in fields, growing in sandy or clay soil. It is particularly abundant in Utah.

CRIMSON MONKEYFLOWER (Mimulus cardinalis)
Snapdragon family *(Scrophulariaceae)*. Blooms red, April-October.
Utah to Oregon, south to northwestern Mexico, 1,800-8,000 feet.

The crimson monkeyflower is a large and conspicuous flower growing along streams and by springs, especially in shady canyons. It reaches a height of from 2-4 feet. Along the Narrows Trail and in the "hanging gardens" on the cliffs of Zion Canyon it may be seen. At Grand Canyon it grows in damp side canyons and near waterfalls.

The yellow monkeyflower is shown on page 73.

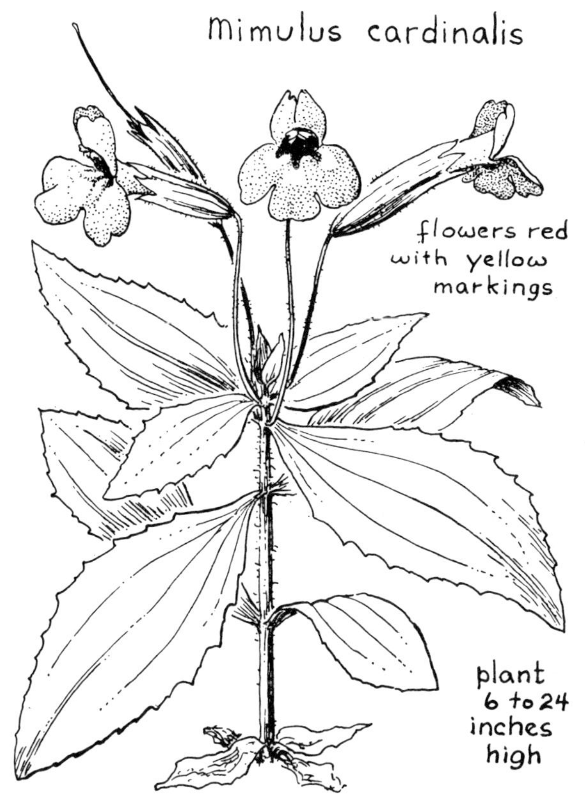

INDIAN PAINTBRUSH, PAINTED CUP (Castilleja)
Snapdragon family *(Scrophulariaceae)*.

Strikingly beautiful plants with large, leafy, petal-like colored bracts which surround the inconspicuous flowers in a spike at the top of the stems. Often partly root parasites. Used medicinally and in ceremonies by the Hopi.

Three common species are illustrated on page 52, easily distinguished by leaf characteristics.

WHOLELEAF PAINTED CUP (Castilleja integra)
Blooms crimson or rose, March-September.
Colorado, New Mexico, Arizona and northern Mexico, 3,000-7,500 feet.

Leaves not toothed, thick and rather narrow but not as narrow as

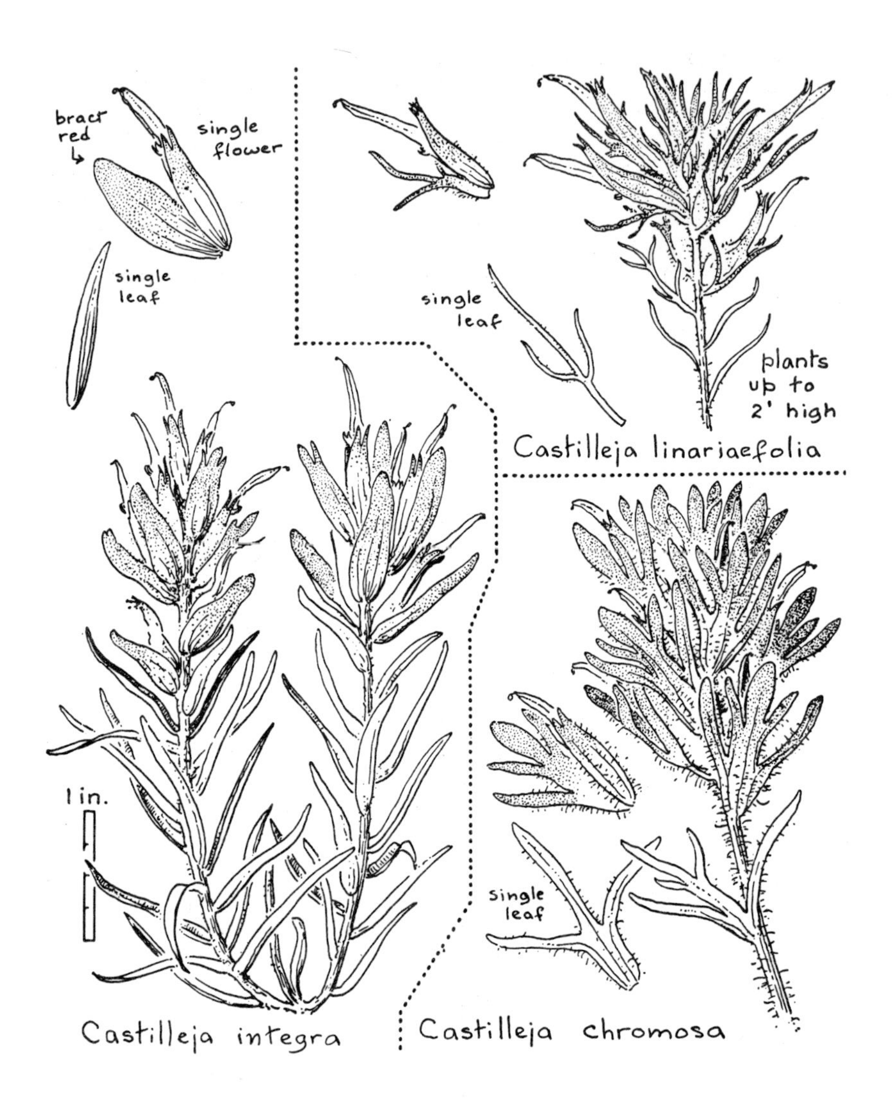

Castilleja linariaefolia leaves. Only slightly hairy. Upper leaf surface smooth. Low growing plant. Stem stout.

WYOMING PAINTED CUP, NARROW-LEAVED PAINTBRUSH (Castilleja linariaefolia)
Blooms yellowish-red to red, sometimes pale, May-September.
Wyoming to Arizona, California and Mexico, 5,000-10,000 feet.

Green tip of flower protrudes from red calyx and bracts. Leaves very narrow, upper ones three-parted. It is reported that the Hopi ate the flowers.

BRIGHT PAINTBRUSH (Castilleja chromosa)
Blooms salmon, light red to rose, March-June

Leaves light grayish-green, rather broad, ends divided into three lobes. Stems and leaves silvery-hairy

MOUNTAIN SNOWBERRY, WOLFBERRY (Symphoricarpos oreophilus)

Honeysuckle family *(Caprifoliaceae)*. Blooms pink, May-August.

Colorado and western Texas to eastern Nevada, Arizona and northern Sonora, 5,500-9,000 feet.

An ornamental shrub belonging to the same family as the well known honeysuckle, it grows on both rims of Grand Canyon and along the upper parts of trails leading into the canyon. At Mesa Verde it frequents the top of Navajo Hill and the bottom of Spruce Tree Canyon. It is fairly common in the Santa Fe and Las Vegas mountains in northern New Mexico.

The thin rounded leaves are smooth and light bluish-green. They are opposite one another on the branch. Livestock and deer browse them.

The tubular flowers, about ½ inch long, are flesh colored or pink and occur in pairs. The buds are purplish-pink. The berry contains two bony seeds and is relished by birds.

The snowberry is sometimes cultivated as an ornamental because of the waxy, snowy-white berries which remain on the bush for a long time.

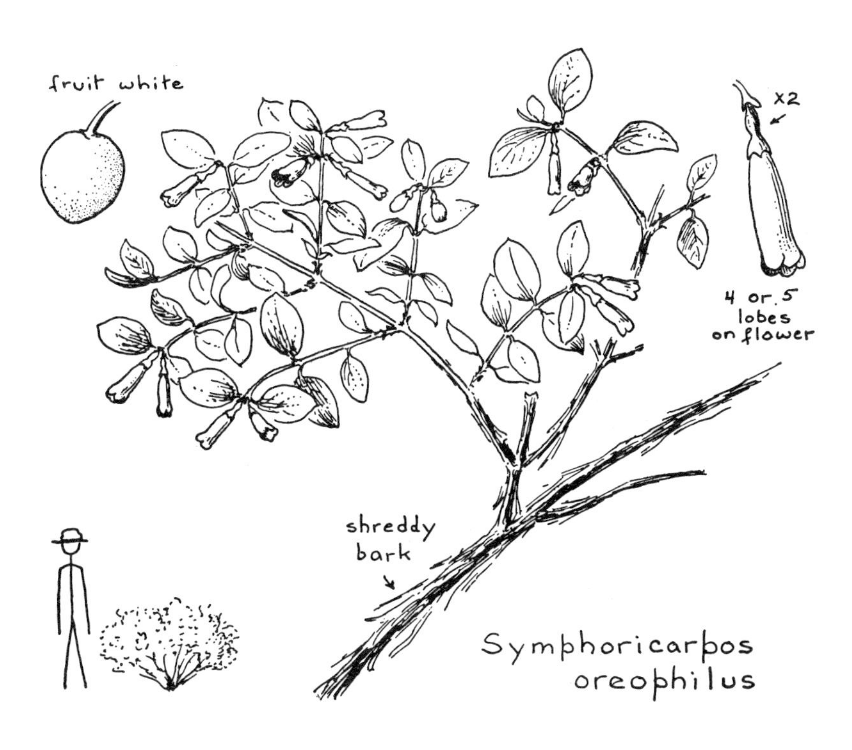

CARDINALFLOWER, LOBELIA (Lobelia cardinalis)
Bellflower family *(Campanulaceae)*. Blooms deep red, July-October.
Widely distributed in the U. S., Mexico and Central America, 4,000-7,000 feet.

The beautiful, tall cardinalflower may be found in moist soil around pools or along streams. At Grand Canyon it grows along Bright Angel Creek, in the lower part of Hance Canyon, and at Toroweap Point. In Zion Canyon it is quite common, being particularly large and abundant in the Court of the Patriarchs.

The leafy stems, often as much as 3 feet high, terminate in a long spike bearing many deep-red, oddly-shaped flowers. It is easy to distinguish, since it is the only conspicuous red flower blooming in late fall.

Some species of lobelia are cultivated as ornamentals.

WIRE LETTUCE, DESERT PINK, SKELETON WEED (Stephanomeria tenuifolia)

Sunflower family *(Compositae)*. Blooms pink, July-August.

Montana to Washington, south to Colorado, northern Arizona and California, 5,000-8,000 feet.

A pale green, sparingly branched, slender plant reaching a height of about a foot and having pretty pink flowers that resemble small carnations. Branches are smooth and rigid and grow at right angles to the main stem.

Leaves on the upper part of the stem are smaller than those towards the base of the plant which are sometimes only tiny scales. The petal-like rays have five small teeth at their tips. The down on the seeds is bright white.

Desert pink grows in Grand Canyon National Park and Navajo National Monument and is also common in the region around Santa Fe.

A closely related species *(Stephanomeria pauciflora)* is native to Grand Canyon and Mesa Verde National Parks and in Wupatki and Navajo National Mounments. Seeds of this desert pink are not as downy as the one described above, and the down is tinged with brown.

It is reported that the Hopi use this species both externally and internally to stimulate milk flow in women.

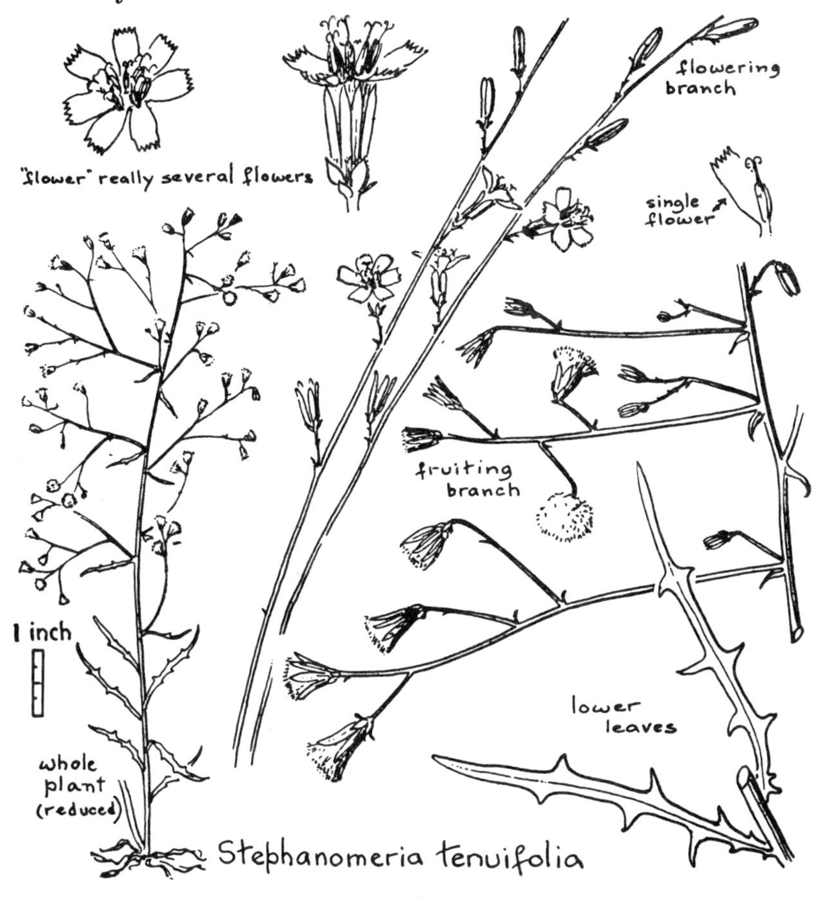

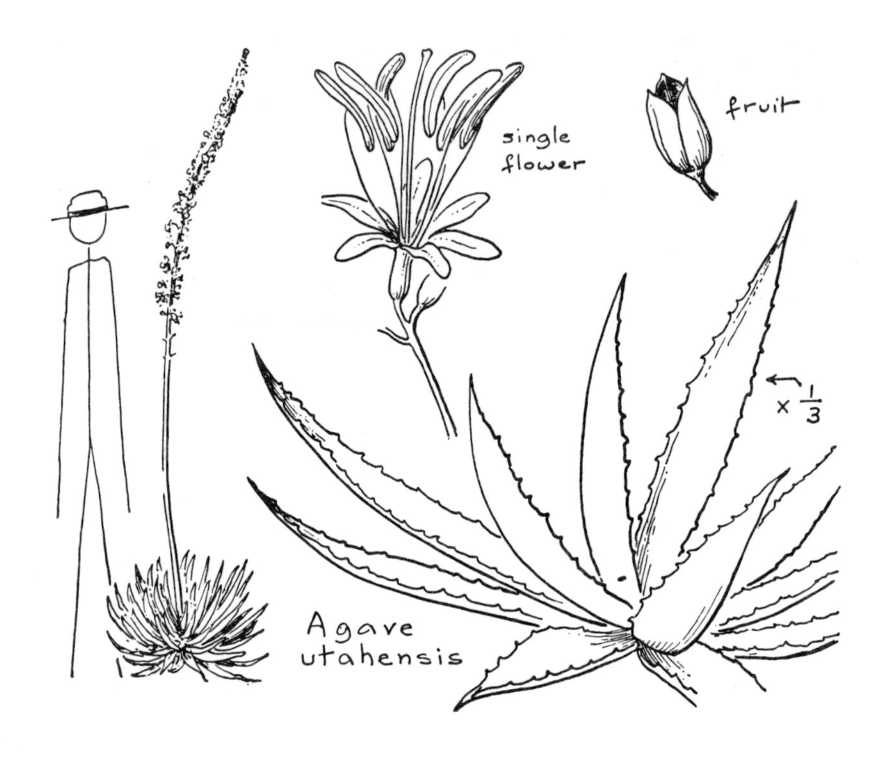

UTAH AGAVE, CENTURY PLANT, MESCAL, (Agave utahensis)
Amaryllis family *(Amaryllidaceae)*. Blooms yellow, May and June.
Utah, Nevada, northern Arizona and northern New Mexico to southeastern California, 3,000-7,500 feet.

The agave is a plant well adapted to arid conditions: (1) the thick leaves serve as water reservoirs; (2) they are covered with a hard cuticle coated with wax to prevent loss of water through evaporation; and (3) the leaves overlap in a rosette arrangement, which itself is a protection against intense light and drouth.

Agaves are sometimes confused with the yucca because of the similar growth habits, but may be distinguished from it, having hooked spines along the leaf margins.

This agave lives about 8 or 10 years, blooms only once, then withers and dies. All of the strength of the plant goes into the production of flowers and fruit. Some southern desert species of agave live 30 years or longer, hence the name "centuryplant."

The fleshy, yellow flowers are borne on a tall spike, which is often curved by the weight of the flowers but which remains standing for a long time after the plant is dead.

An intoxicating drink, "tequila," is made from the agave, and Indians eat the roasted bud of the flower stalk (mescal).

The illustrated species of agave grows at Grand Canyon on both rims and in the canyon. It is conspicuous on the Tonto Plateau where it grows associated with blackbrush *(Coleogyne ramosissima)*. Old Indian mescal roasting pits are found in Grand Canyon.

Mountain-mahogany

Ceanothus in bloom

Mallow, a "burning bush"

Mormon-tea, Ephedra

Cliffrose

Senecio, Groundsel

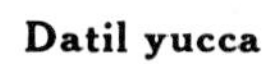

Datil yucca

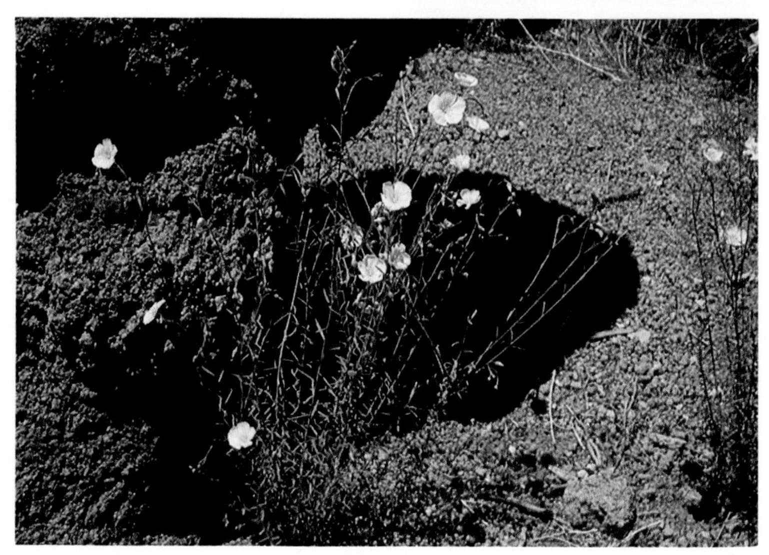

Blue flax

SULPHURFLOWER, SULPHUR ERIOGONUM (Eriogonum cognatum)
Buckwheat family *(Polygonaceae)*. Blooms yellow, July-September.
Arizona, 5,000-7,000 feet.

This species is very much like two other well known sulphur-flowers *(Eriogonum stellatum)* and *(Eriogonum umbellatum)*. These are handsome plants with many bright yellow flowers which turn reddish in the fall. The leaves are woolly.

Plants are common and occur in Grand Canyon and Mesa Verde National Parks, Aztec Ruins and Arches National Monuments, and also in Betatakin Canyon in Navajo National Monument.

DESERTTRUMPET, BLADDERSTEM (Eriogonum inflatum)
Blooms yellow, fading pinkish, March-July.
Utah and Arizona to southern California and Baja California, around 4,000 feet.

This is a weird looking plant with hollow, inflated, pale bluish-green stalks and a few tiny yellow, silky flowers which turn pinkish as they fade. It grows about 1-2 feet high. The leaves are darker green and grow at the base of the plant.

Pinkish erigonums on page 38.

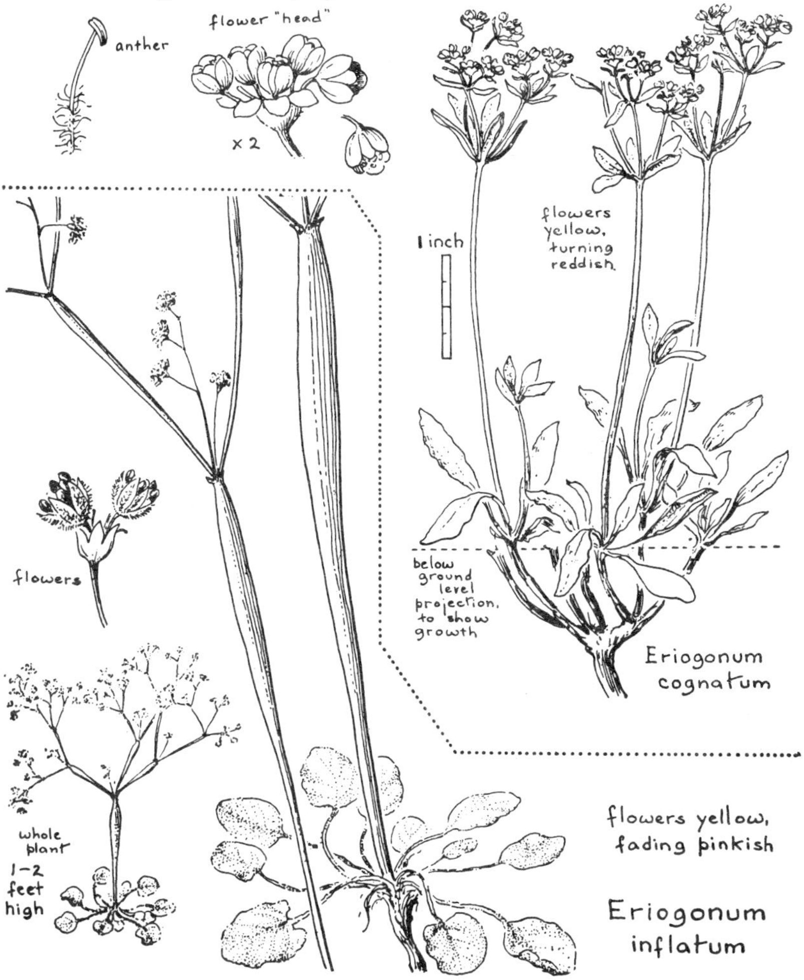

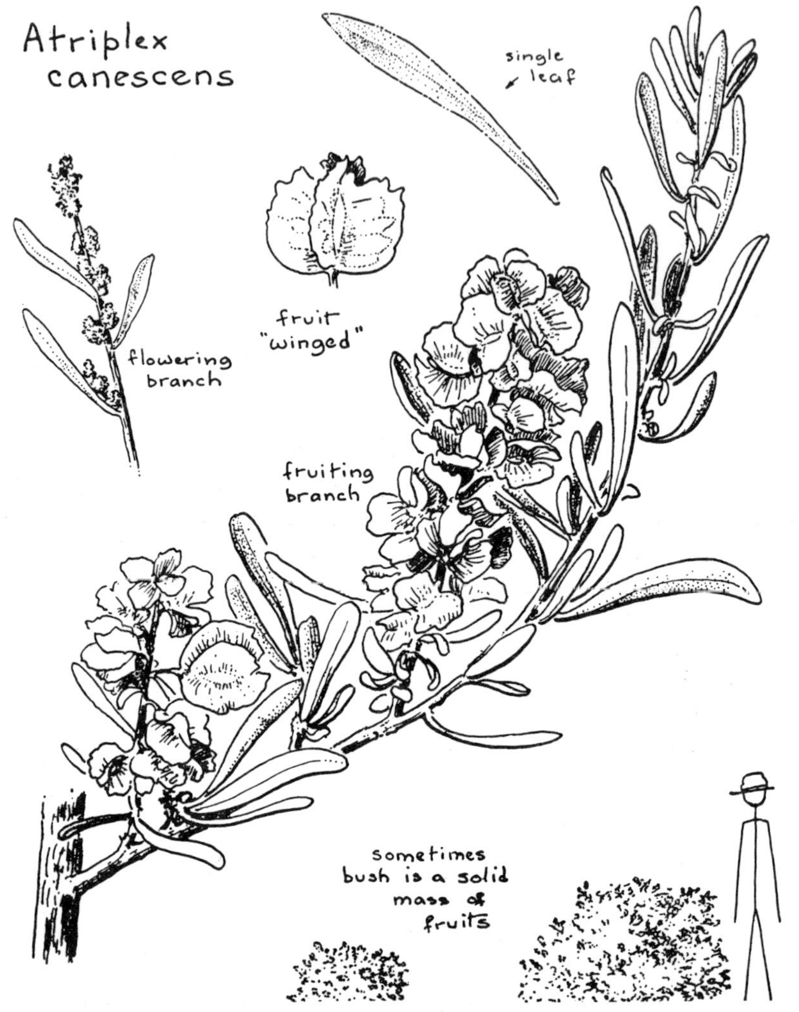

FOURWING SALTBUSH, SHAD-SCALE, CHAMISO (Atriplex canescens)
Goosefoot family *(Chenopodiaceae)*. Blooms pale yellow, late summer and fall.
South Dakota to Oregon, south to northern Mexico, 6,500 feet or lower.

One of the commonest shrubs of the Southwest is the saltbush. It is adapted to diverse soil and climatic conditions and grows from the southern desert up to the Ponderosa Pine Belt in alkaline and non-alkaline soil. It is a valuable nutritious browse plant.

The bush is woody, rigid and freely branched, with grayish or yellowish-green, narrow leaves. It grows about 3 feet high, and has inconspicuous flowers. The four-winged fruit is prominent, appearing in large bunches.

YELLOW COLUMBINE (Aquilegia chrysantha)
Buttercup family *(Ranunculaceae)*. Blooms yellow, April-September.
Southern Colorado to western Texas, Arizona and northern Mexico, 3,500-11,000 feet.

This large, canary-yellow, long-spurred columbine ranks as one of the most beautiful wild flowers of this region. It likes damp, rich soil, and has an exceptionally wide altitudinal range.

The nectar in columbine flowers forms in the spurs of the petals and can usually be reached only by "long-tongued" insects and hummingbirds. However, bees get the nectar by cutting holes in the spurs.

Yellow columbine grows in the "hanging gardens" in Zion Canyon, and at Ribbon Falls in Grand Canyon.

Red columbine is described on page 40.

YELLOW BEEPLANT, YELLOW SPIDERFLOWER (Cleome lutea)

Caper family *(Capparidaceae)*. Blooms light yellow, May- July.

Nebraska to Washington, south to New Mexico, Arizona and eastern California, 4,800-6,000 feet.

The yellow beeplant is less common than the purple one (see page 95). Pickled flower buds of a European relative of the beeplant are the "capers" used in seasoning. The Hopi use the young plants as pot herbs.

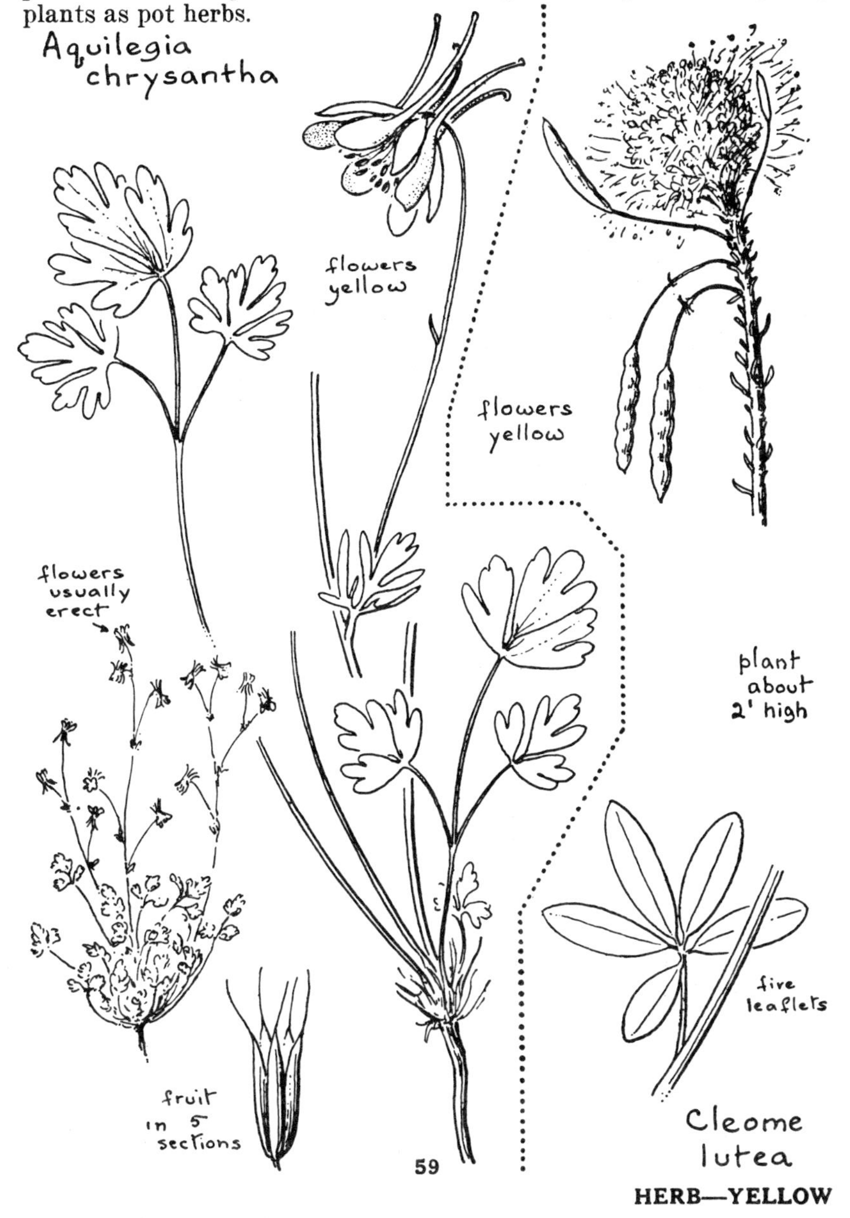

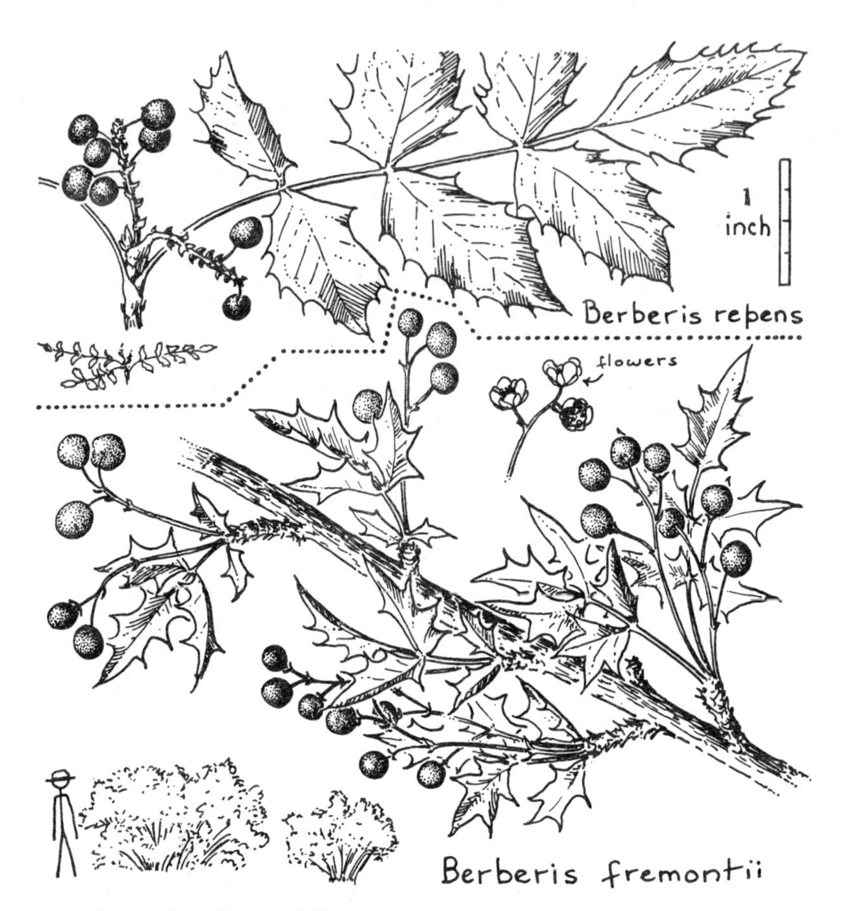

HOLLYGRAPE, FREMONT BARBERRY (Berberis fremontii)
Barberry family *(Berberidaceae)*. Blooms yellow, May-July.
Colorado and Utah to New Mexico and northern Arizona, 4,000-7,000 feet.

Hollygrape is an appropriate name for this shrub, as it has a leaf like a holly and fruit like a grape. It grows within the Pinyon-Juniper Belt, reaching a height of from 6 to 8 feet. At Grand Canyon it resembles a small tree rather than a shrub.

The evergreen leaves are compound; the five, spiny-toothed leaflets small, thick and leathery. The stems are spineless and the six-parted yellow flowers are arranged in short, loose clusters. Berries are dark blue, dry, and inflated at maturity.

The Hopi make various articles with the yellow wood of hollygrape. The plant contains berberine, a drug, and Indians use the root in making a tonic and a yellow dye.

A close relative of hollygrape is the oregongrape *(Berberis repens)* which has a much wider climatic range than does the hollygrape. The creeping root stock of this low-growing shrub makes an excellent ground cover, protective against erosion.

The yellow blooms come in April and May; flower clusters are few, long, many-flowered and fragrant. The small, blue-black or purple berries are juicy and good for making jelly, and are eaten by birds and various mammals. The leaves turn red in autumn.

The Navajo makes a medicine for rheumatism from the leaves and stems of oregongrape.

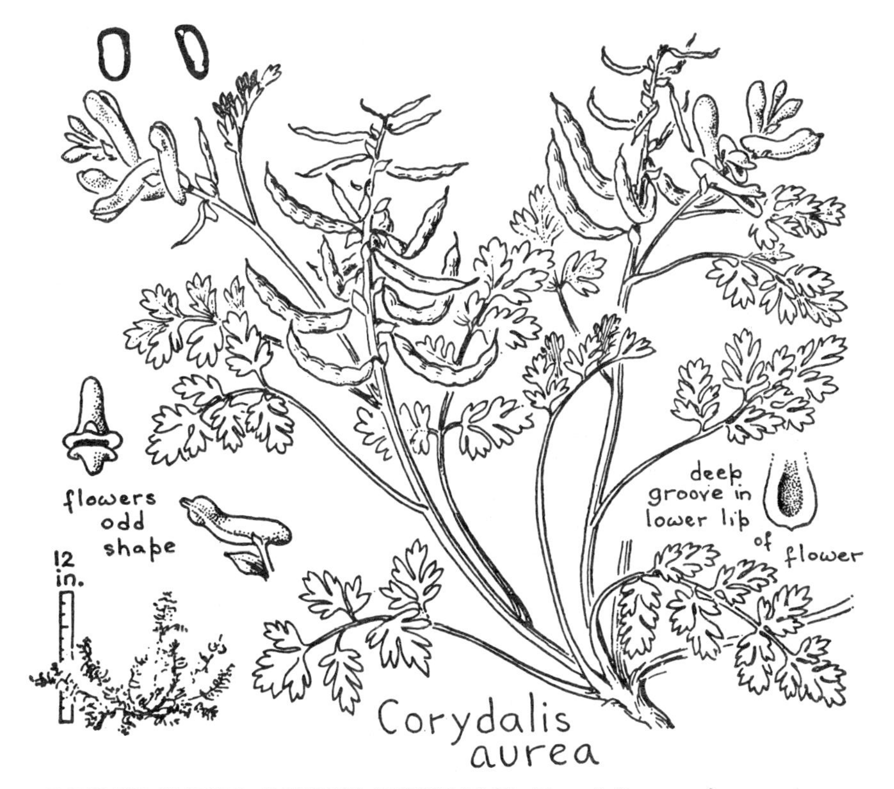

GOLDEN SMOKE, GOLDEN CORYDALIS (Corydalis aurea)

Poppy family *(Papaveraceae)*. Blooms yellow, February-June.

Nova Scotia to Alaska, south to Pennsylvania, Arizona, northern Mexico, and California, 2,500-8,000 feet.

Golden smoke is easy to recognize with its silvery blue-green, feathery foliage and its strangely-shaped, irregular, golden flowers. It is a close relative of the eastern wild bleedingheart and Dutchmans-breeches.

This plant has rather weak, slender stems from 6-14 inches high. It does not stand erect—some branches lie on the ground. The stems are hollow and contain a watery juice which is said to be poisonous to sheep, if eaten freely.

The flowers remind one of little birds. They grow clustered together on a spike, their heads turning in different directions. There are two little pointed green sepals at the base of the flower. Usually only one of the outer petals is spurred; the small inner petals are united at the top enclosing the inner flower parts.

The seed pods are curved and come out from the spike at different angles and each pod contains numerous black shiny seeds.

Golden smoke has a wide altitudinal range and may be found in damp protected places throughout the Southwest.

This plant is listed from Grand Canyon and Mesa Verde National Parks and in Aztec Ruins, Walnut Canyon, Navajo and Bandelier National Monuments. It also grows in the hills near Santa Fe.

A similar but longer spurred species *(Corydalis montana)* is found in Zion National Park.

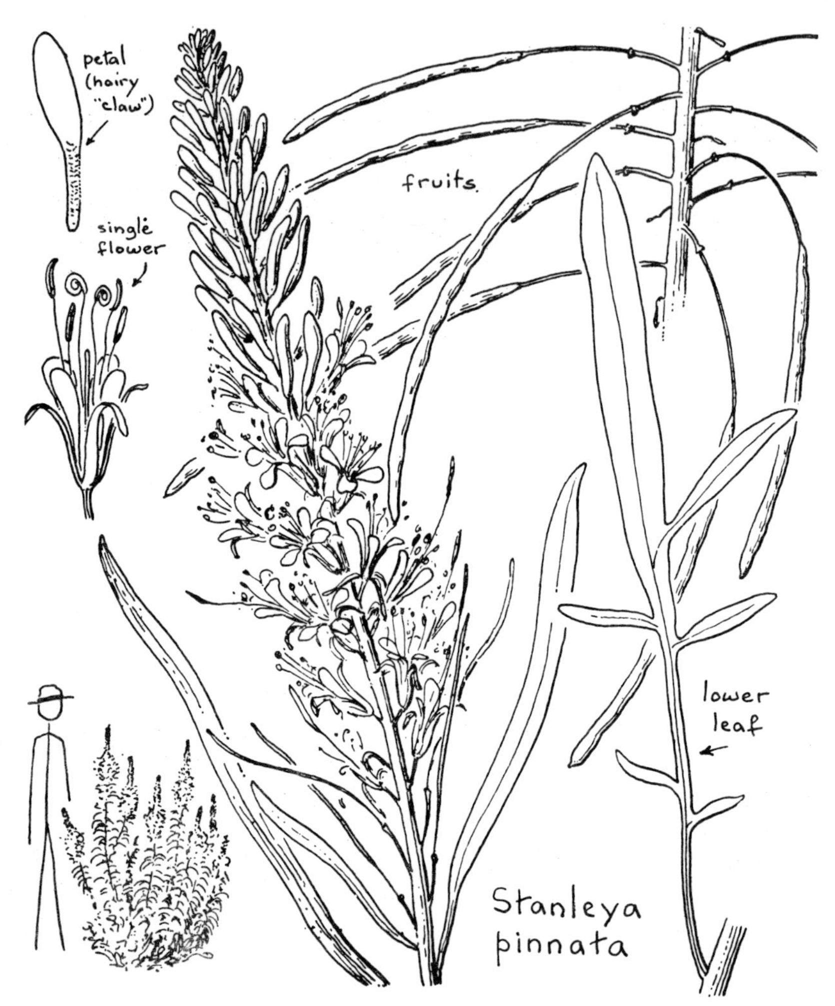

DESERTPLUME, PRINCES PLUME (Stanleya pinnata)
Mustard family *(Cruciferae)*. Blooms yellow, May-July.
North Dakota to Idaho south to Texas, Arizona and California, 2,500-6,000 feet.

This plant, with the golden plumes showing above the semidesert chaparral, grows not only on dry mesas and plains with sagebrush and pinyons, but also at lower elevations. It has tall stout stems, rather woody at the base, and is tolerant of soil containing gypsum.

The pale green, alternate leaves are narrow and simple, although some at the base are divided. The flower spike usually shows crowded buds at the top, flowers farther down, and the long, narrow drooping seed pods at the base. Flowers have four sepals and four petals.

Desertplume is used as a pot herb by Indians, who also make a mush with the seeds.

The species is found in National Parks and Monuments as follows: Grand Canyon, Zion, Mesa Verde, Aztec Ruins, Chaco Canyon, Wupatki, Pipe Springs, Colorado and Petrified Forest.

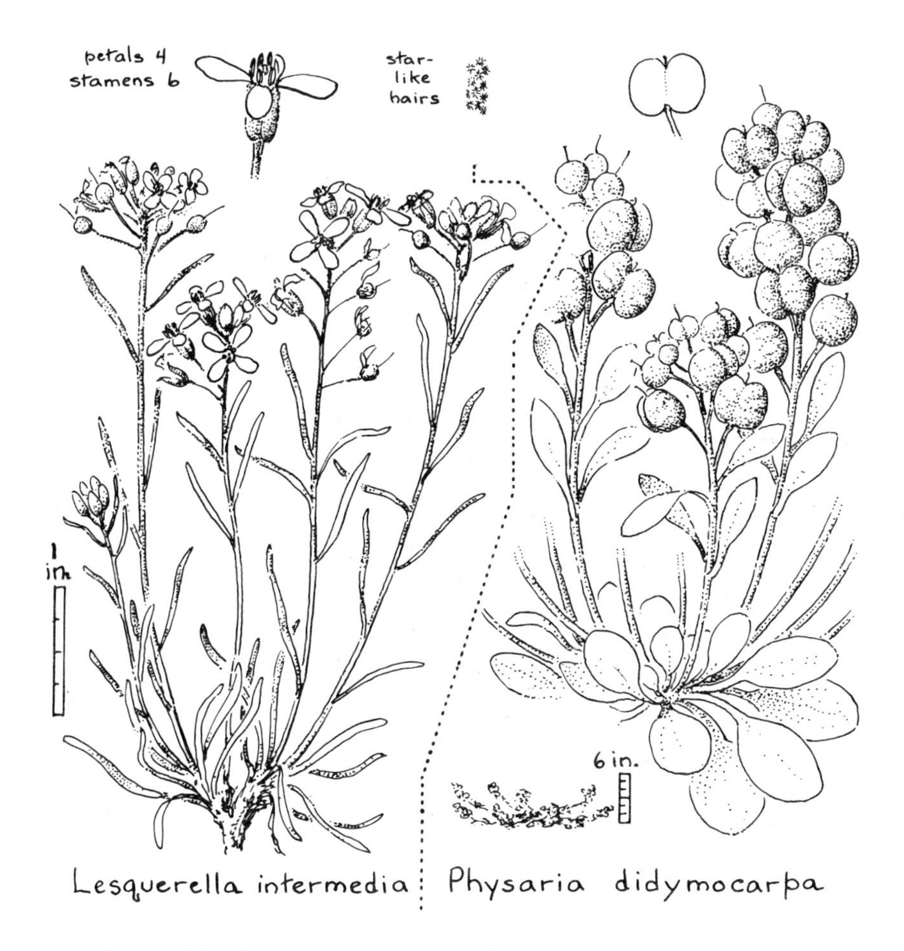

BLADDERPOD (Lesquerella intermedia)
Mustard family *(Cruciferae)*. Blooms yellow fading reddish, April-August. Colorado, Utah, New Mexico and Arizona, 5,500-7,200 feet.

This is a common perennial less than 1 foot high with yellow flowers that sometimes fade reddish. It is a hairy plant, the hairs star-like.

Roots of this species are used by the Hopi as an antidote for rattlesnake bite.

May be seen at Yavapai Point on the south rim of Grand Canyon and also at Walnut Canyon and Navajo National Monuments, and Petrified Forest National Park.

TWINPOD, DOUBLE BLADDERPOD (Physaria didymocarpa)
Same family. Blooms pale yellow, May-July.
Saskatchewan and Alberta to Colorado and Nevada, 5,500-6,500 feet.

An odd little gray-green plant only about 6 inches high, with showy, pale yellow petals and rounded, inflated, thin papery-walled pods with two divisions; hence the name twinpod.

The leaves form a rosette at the base of the plant and the hairs are also star-like. Many of the stems are prostrate.

It grows in dry places in Zion and Mesa Verde National Parks.

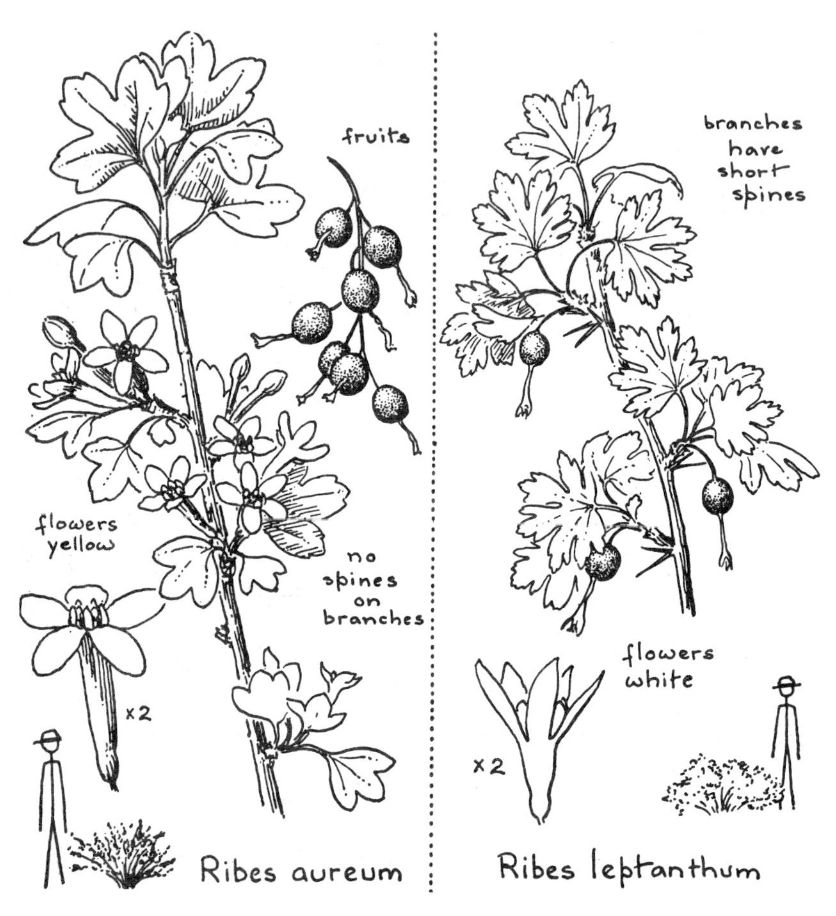

GOLDEN CURRANT, BUFFALO CURRANT (Ribes aureum)
Saxifrage family *(Saxifragaceae)*. Blooms golden yellow, March-June.
South Dakota to southern Saskatchewan and Washington, south to New Mexico, Arizona and California, 5,000-6,000 feet.

The golden currant is one of the most showy shrubs in the Southwest, growing beside streams and in moist canyons, and giving the effect of forsythia when it is in bloom with its long, tubular, fragrant golden flowers. It is often cultivated as an ornamental.

There are no spines. The leaves are bright, glossy green; the smooth woody stems, pinkish-gray. The edible fruit is large, yellowish red or black.

TRUMPET GOOSEBERRY (Ribes leptanthum)
Blooms white or yellowish, May-June.
Colorado, Utah, New Mexico and Arizona, 6,000-9,500 feet.

The trumpet gooseberry is a freely branched spiny shrub with rigid stems, growing along streams. The berry is black or dark red and though very tart was eaten fresh or dried by Indians. The fruits are also relished by birds, and the plant is browsed by domestic animals and deer.

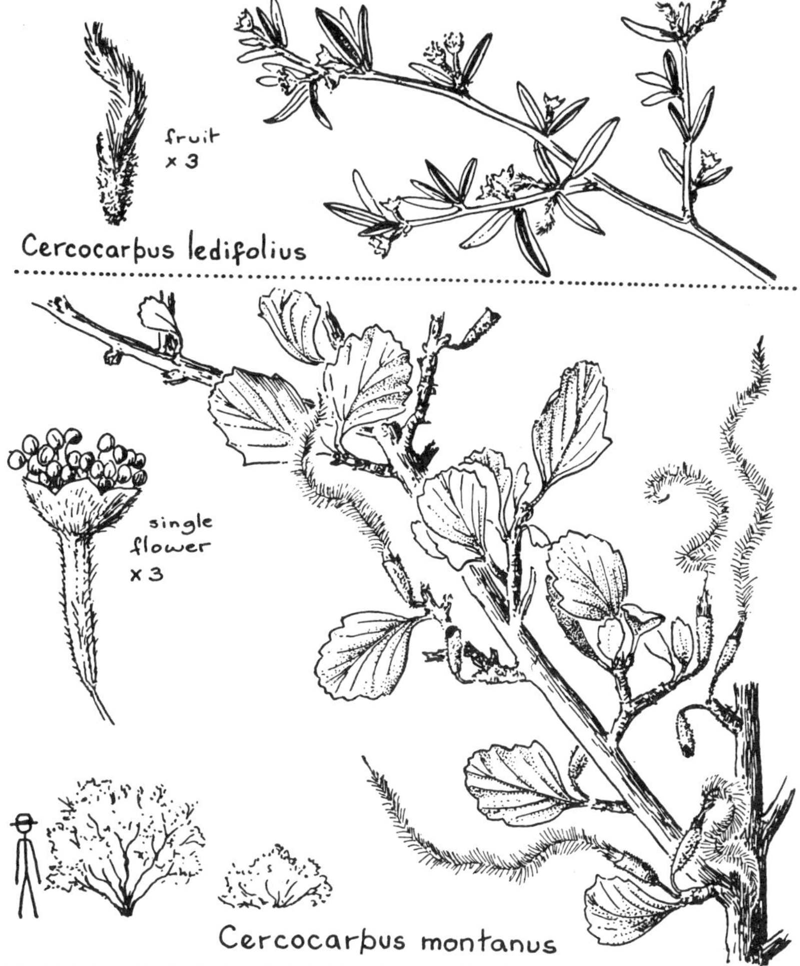

CERCOCARPUS, MOUNTAIN-MAHOGANY (Cercocarpus)

Rose family *(Rosaceae)*. Blooms yellowish or dull white, early spring.

A widely branching shrub or small tree with stout limbs, it is helpful in preventing erosion and is valuable as a browse plant. The hard wood is useful and a red-brown dye can be made from the bark.

The flowers are small and inconspicuous and petals are absent. The seed has a long, spiral plume.

Two common species are illustrated here.

ALDERLEAF MOUNTAIN-MAHOGANY, DEERBROWSE (Cercocarpus montanus)

South Dakota and Montana to Kansas, New Mexico and Arizona, 4,500-6,800 feet.

Grows in canyons and on hills of sagebrush, pinyon and ponderosa pine belts. Occurs in Grand Canyon, Zion and Mesa Verde National Parks and in Bandelier, Chaco Canyon and Canyon de Chelly National Monuments.

CURLLEAF MOUNTAIN-MAHOGANY (Cercocarpus ledifolius)

Montana to Washington, south to Colorado. Upper Pinyon, Ponderosa Pine and Aspen belts.

In central Nevada it replaces the ponderosa pine in the Transition Zone. It occurs in Grand Canyon and Zion National Parks.

The leathery leaves are evergreen, rather narrow, and pointed at both ends, the edges slightly curled under.

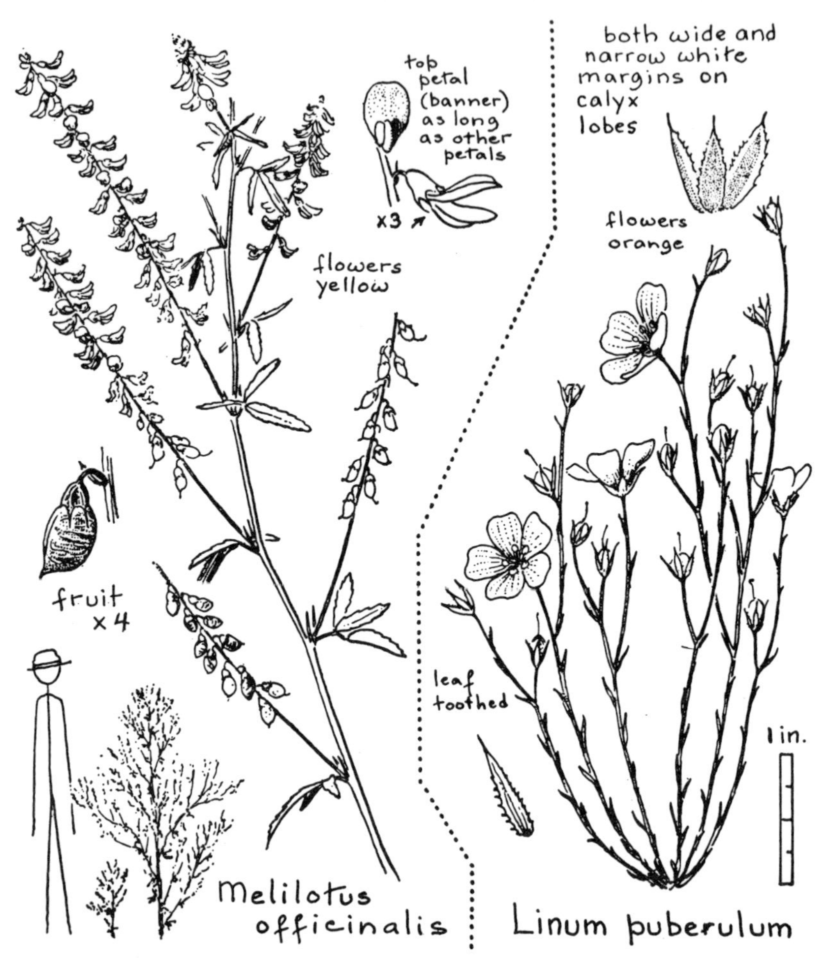

YELLOW SWEETCLOVER (Melilotus officinalis)
Pea family *(Leguminosae)*. Blooms yellow, July to October.

There are two common sweetclovers, yellow sweetclover *(Melilotus officinalis)* and white sweetclover *(Melilotus albus)*. See page 28. Both are tall, fragrant, roadside weeds introduced from Europe and distributed throughout the United States. They often grow in solid masses for many miles.

These are excellent honey plants, blooming nearly all summer. They can tolerate alkaline soil.

YELLOW FLAX (Linum puberulum)
Flax family *(Linaceae)*. Blooms yellow tinged with orange, April-July. Colorado to Utah, New Mexico and Arizona, 3,500-6,500 feet.

A small flax with yellow flowers tinged with orange or coral, darker towards the center, its petals are prettily veined.

It is common on the dry hills around Santa Fe and may be seen at Grand Canyon National Park and at Walnut Canyon, Aztec Ruins, Bandelier, Chaco Canyon and Montezuma Castle National Monuments.

The blue flax is described on page 99.

Both species of flax have delicate petals that fall readily, but the sepals remain in place.

The native species are not of commercial importance, but their

Old World relative *(Linum usitatissimum)*, is used in the production of linen from the fibrous stems and linseed oil from the seeds. It is reported that some western Indians used the long stem fibers of flax to make cord.

SQUAWBUSH, SKUNKBUSH (Rhus trilobata)

Sumac or Cashew family *(Anacardiaceae)*. Blooms yellow, March-June. Saskatchewan to Washington, south to Mexico, 2,500-7,500 feet.

Squawbush is one of the most common shrubs of the pinyon-juniper woodland. It occurs in all of the National Parks and Monuments in which this zone is represented, and also in the southern desert.

A woody shrub with wide, irregular branching, and dark reddish-brown bark, it varies from 1-8 feet high, but is usually less than 6 feet. The shiny deciduous leaves are divided into three leaflets which vary in size and shape with the individual plants. They have a strong odor; hence the name skunkbush.

The plant is not poisonous but is closely related to the common poison ivy *(Rhus radicans)*. The leaves, which are browsed by a few animals, turn a beautiful rich red in the fall.

The five-parted flowers are yellow and grow in dense clusters on a spike, usually appearing before the leaves. The fruit is a sticky, bright, orange-red berry, pleasantly acid to taste. A lemonade-like drink may be made from the berries and Indians used them for food and in making dye. Birds and wild animals also eat the berries.

Mexicans and Indians use the twigs of squawbush in basketry.

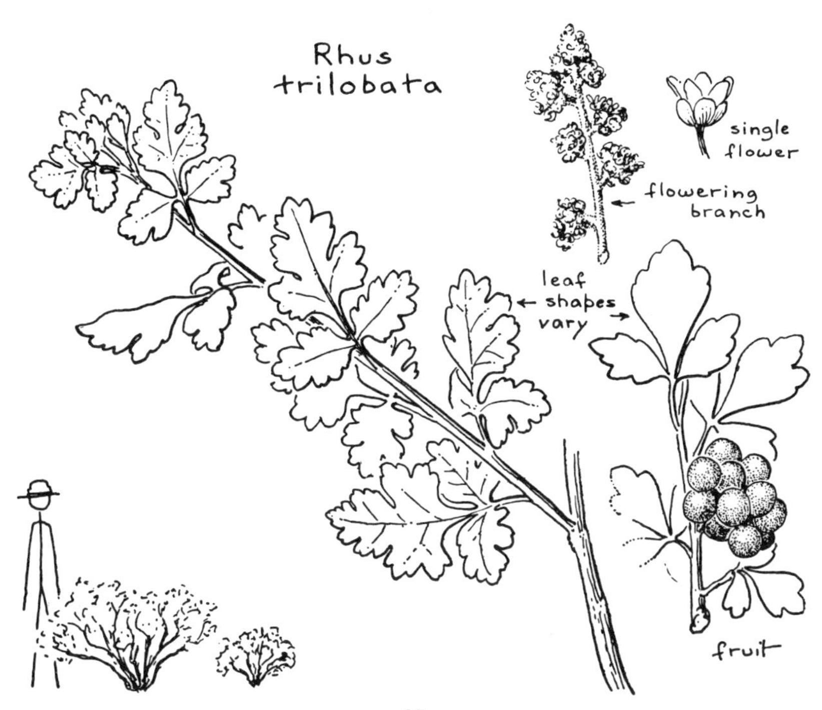

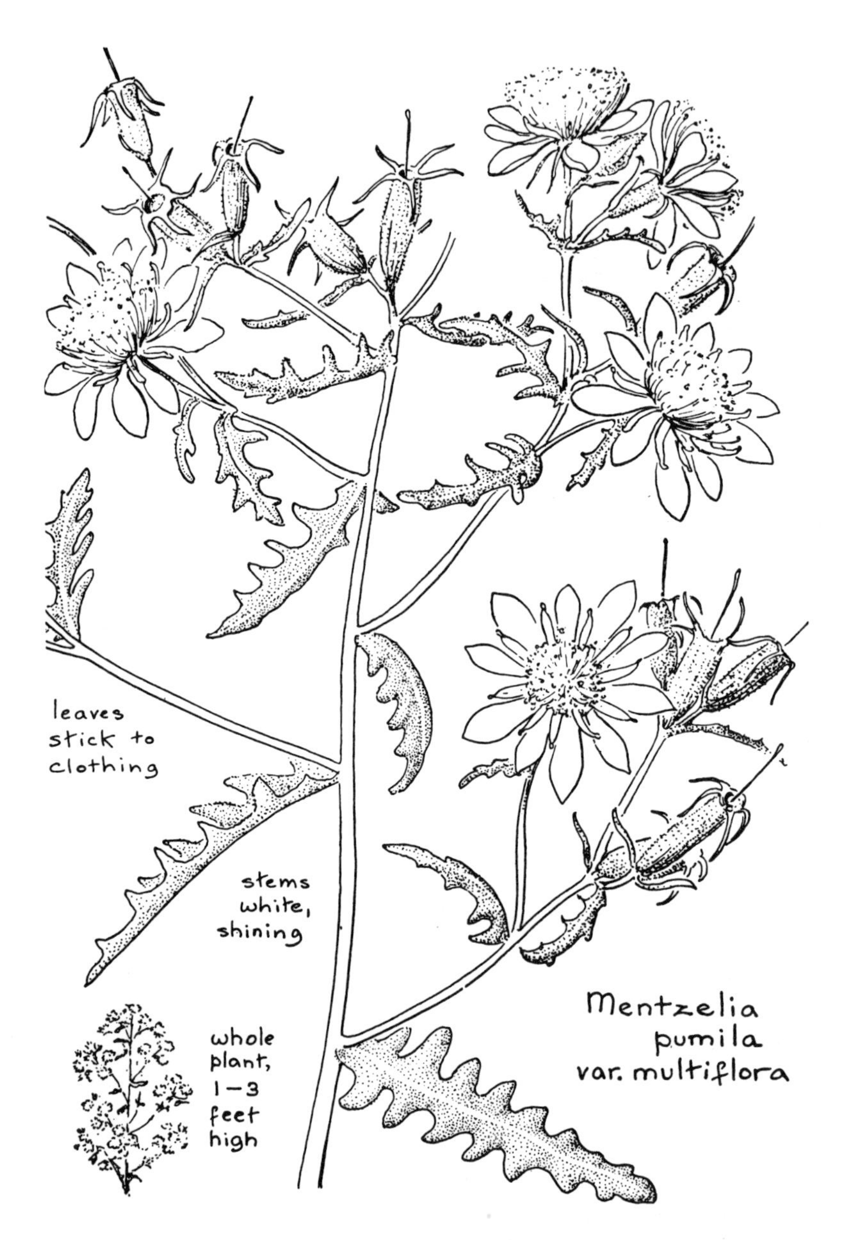

MENTZELIA, EVENING-STAR, STICKLEAF (Mentzelia pumila var. multiflora)

Loasa family *(Loasaceae)*. Blooms yellow or cream, May-August.
Colorado to New Mexico and Arizona, 100-7,000 feet.

Occurs in Grand Canyon, Zion, Petrified Forest, and Mesa Verde National Parks and in Aztec Ruins, Wupatki, Bandelier and Montezuma Castle National Monuments.

The star-shaped flowers have ten petals and open in the late afternoon. It also has many stamens of which a few outer ones are petal-like. Stems are freely branching, and white, and the leaves are light green and sticky.

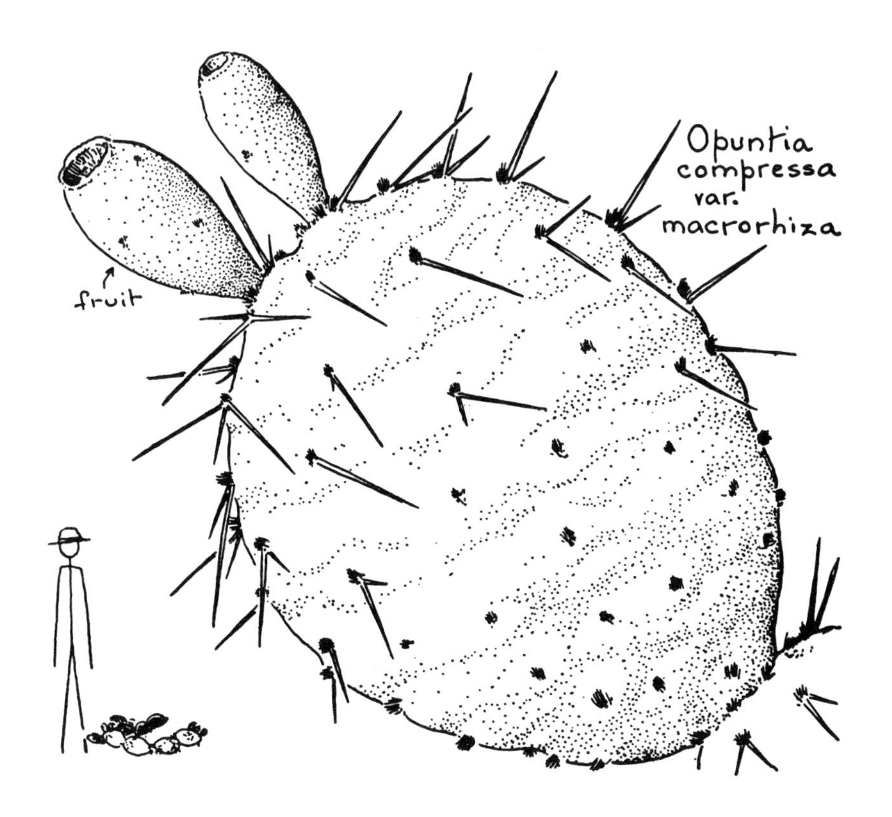

TUBEROUS PRICKLYPEAR, YELLOW-FLOWERED PRICKLYPEAR
(Opuntia compressa var. macrorhiza)

Cactus family *(Cactaceae)*. Blooms yellow to reddish, May-June.
New Mexico and Arizona, 4,500-6,500 feet.

This pricklypear is a creeping plant with a thickened, tuberous root. The lovely flowers are large, with their many waxy petals yellow, orange or reddish towards the base.

The blooms usually last only one day, as is the case with most cactus flowers. The flower cup is filled with many golden stamens that are sensitive and curl towards a bee when it alights in the flower. The cylindrical, greenish-yellow or reddish fruit is fleshy and without long spines.

Joints are wrinkled crosswise. The white spines are stout and rigid and are only two or three to a group, while groups of spines are about 1 inch apart and occur only on the upper half of the joints and along their edges.

The pricklypears spread easily and sometimes become a pest, although they are occasionally used as forage when the spines are burned off.

This species is common in the hills around Santa Fe. It is reported from Canyon de Chelly National Monument.

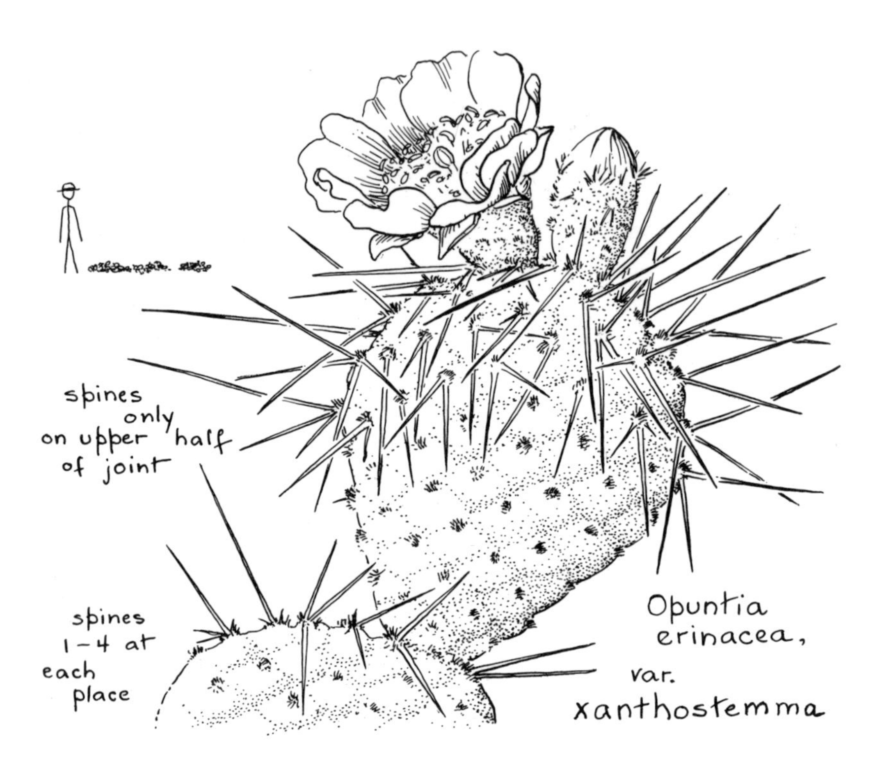

RED-FLOWERED PRICKLYPEAR, HEDGEHOG PRICKLYPEAR (Opuntia erinacea var. xanthostemma) Opuntia erinacea var. utahensis.
Cactus family *(Cactaceae)*. Blooms yellow, apricot, pink or red, May-June. Western Nebraska, Colorado, Utah and Arizona, 4,500-7,200 feet.

Cactus plants are superb examples of plant adaptation to desert conditions. With extensive, shallow roots they are well equipped to take up available moisture, which they store in their fleshy joints. These joints expand and become fat when there is plenty of moisture, and contract, becoming thin and wrinkled in times of drought. A thick skin prevents water evaporation, and spines protect the plant against inroads of thirsty desert rodents.

The pricklypears have flat joints and the long spines are not barbed or covered with a papery sheath as are those of the chollas.

This species of pricklypear has a beautiful, delicate, waxy flower varying in color from yellow to apricot, pink or red. The mature fruit is dry and spiny.

The body of the plant is somewhat lead-colored with edges of the joints darkish. The spines are about 1-1½ inches long, white with darker tips, and grow only on the upper half of the joint.

Found on flats and rocky hills of the Pinyon-Juniper Belt, and also with the ponderosa pine where it is the most abundant cactus, it is very common in Zion, Grand Canyon and Mesa Verde National Parks. It also has been reported at Petrified Forest National Park and Navajo National Monument.

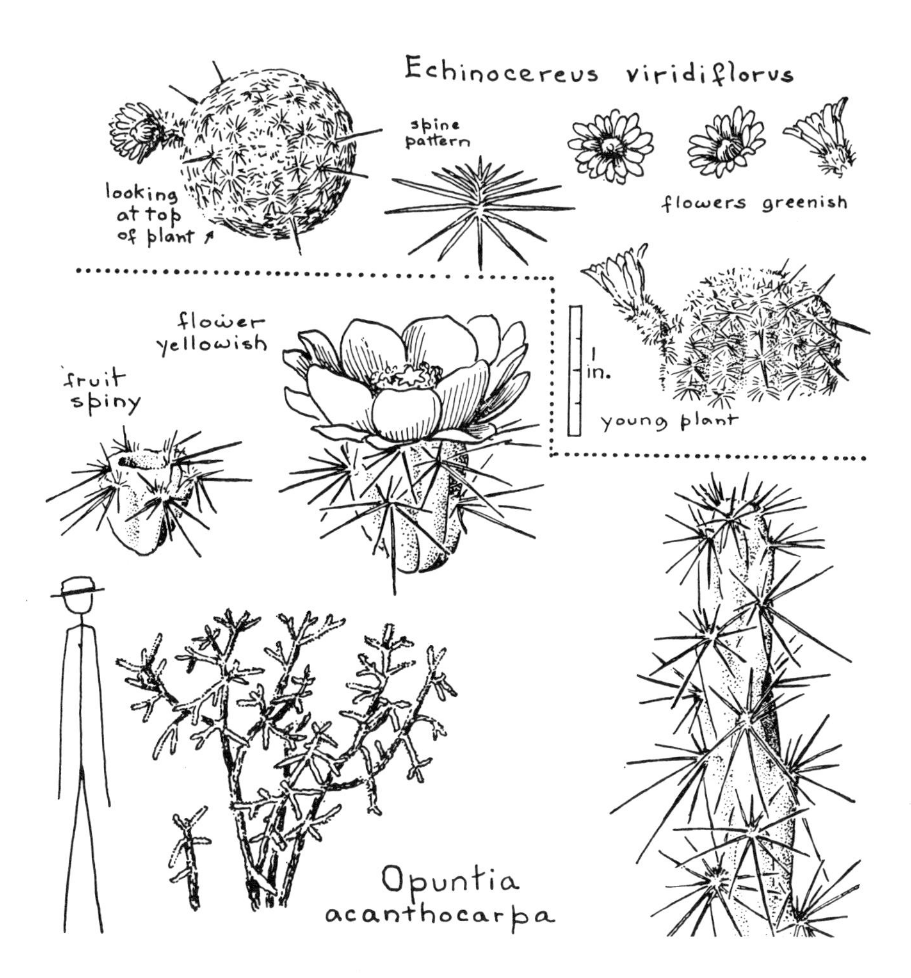

VARIED HEDGEHOG, GREEN-FLOWERED TORCH CACTUS (Echinocereus viridiflorus)

Cactus family *(Cactaceae)*. Blooms yellowish-green, May-June.
Southern Wyoming to New Mexico and western Texas, 7,500 feet and lower.

The spines of this tiny cactus grow on vertical ridges. They are white, dark brown or purple, arranged in light and dark bands around the plant.

BUCKHORN CHOLLA, YELLOW-FLOWERED CANE CACTUS (Opuntia acanthocarpa)

Blooms greenish-yellow, spring.
Southwestern Utah and southern Nevada to Sonora and southern California.

The buckhorn cholla occurs more commonly in the southern desert than it does in the sagebrush of the mesa country, yet there are some places in this region where it is abundant and conspicuous. It grows at Coalpits Wash and in the lower part of Zion Canyon. It occurs between Kanab, Utah, and the Kaibab Plateau and also on the south rim of Grand Canyon.

It is an odd, sprawling cactus with a short trunk or sometimes no trunk at all. The spines are barbed and covered with straw-colored papery sheaths.

The flowers are not conspicuous but are quite pretty. The waxy petals are greenish yellow and stay fresh only about one day. The fruit is dry and spiny.

The Pima Indians use the flower buds for food, steaming them in a pit and removing the spines.

GROUNDCHERRY, STRAWBERRY TOMATO (Physalis fendleri)

Potato family *(Solanaceae)*. Blooms yellowish, May-August.

Colorado and Utah to Arizona, southern California and northern Mexico, 3,300-7,500 feet.

Groundcherry of this species is a widely branched, straggling, rather weedy plant growing about 1 or 2 feet high from a deep tuberous root. It is found on dry mesas and slopes with the pinyon and juniper trees.

As a member of the useful potato family it is related to the tomatoes, red peppers, and tobacco plants, also to such poisonous species as henbane and belladonna, sources of certain drugs.

The pale, dull yellow flower is in a nodding position when it opens. The petals are marked inside toward the center with brown.

When the fruit develops, the thin, papery, fine-toothed calyx becomes enlarged, completely covering the fruit.

The round, yellow, many-seeded berry is edible and is sometimes preserved. Indians ate them raw and cooked.

This species of groundcherry is common in Zion Canyon; at Grand Canyon it may be seen along the Kaibab Trail. It also occurs at Wupatki, Bandèlier and Chaco Canyon National Monuments.

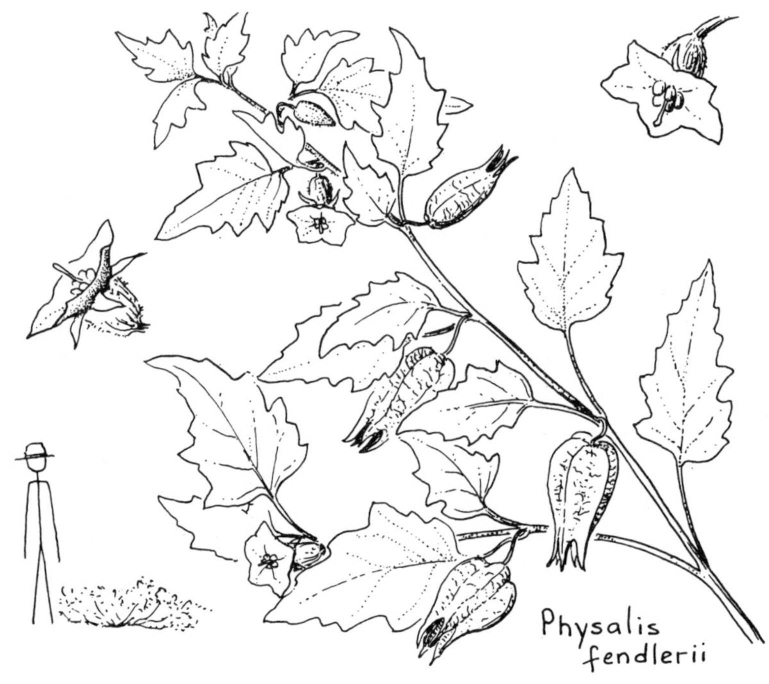

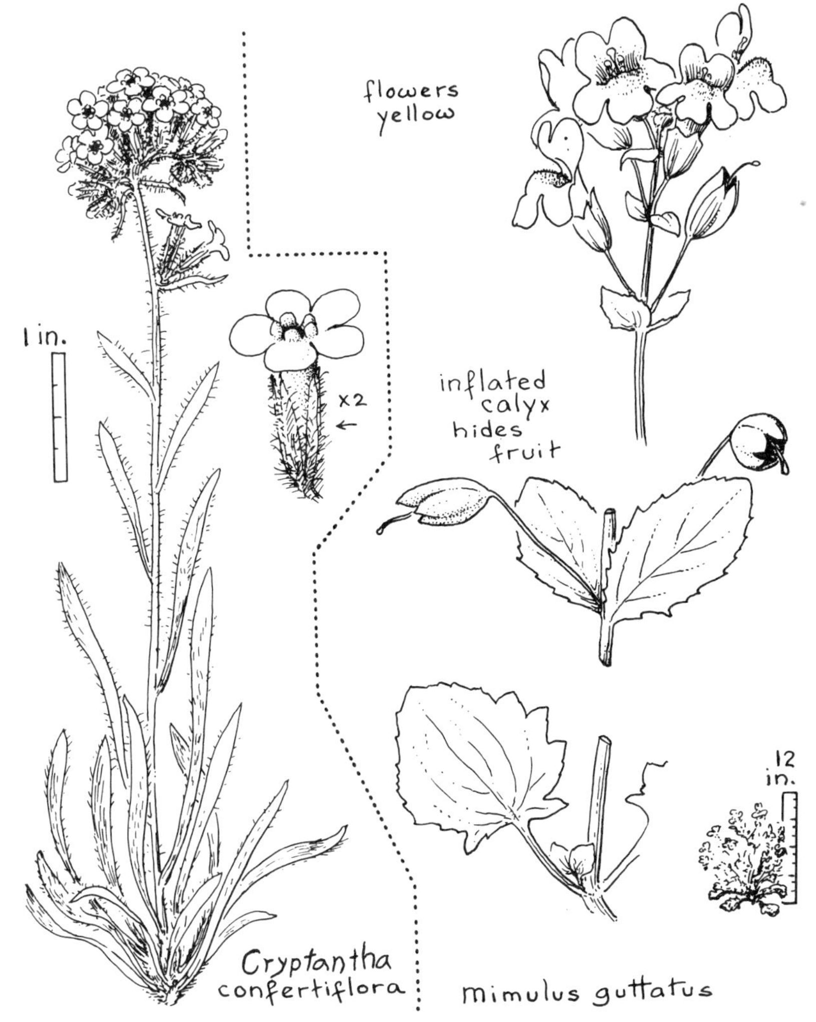

CRYPTANTHA, YELLOW BORAGE (Cryptantha confertiflora)
Borage family *(Boraginaceae)*. Blooms pale yellow, spring and summer. Western Utah to northwestern Arizona and California, around 5,000 feet.

This rather coarse, pale green, hairy little plant is capped with a head of small, light-yellow flowers and grows from thick woody roots.

It is associated with the junipers, commonly on limestone, and seems to prefer dry open spaces. It is found in Zion National Park and Navajo National Monument; at Grand Canyon it is one of 14 species of cryptantha.

YELLOW MONKEYFLOWER (Mimulus guttatus)
Snapdragon family *(Scrophulariaceae)*. Blooms yellow, March-September. Montana to Alaska, south to New Mexico, 1,300-8,000 feet.

The yellow monkeyflower reaches about 1 foot in height. It has dark red spots on the hairy throat of the flower tube. Abundant by springs and along streams, it occurs in Grand Canyon and Zion National Parks with the crimson monkeyflower, and in Montezuma Castle and Bandelier National Monuments.

The red monkeyflower is pictured on page 51.

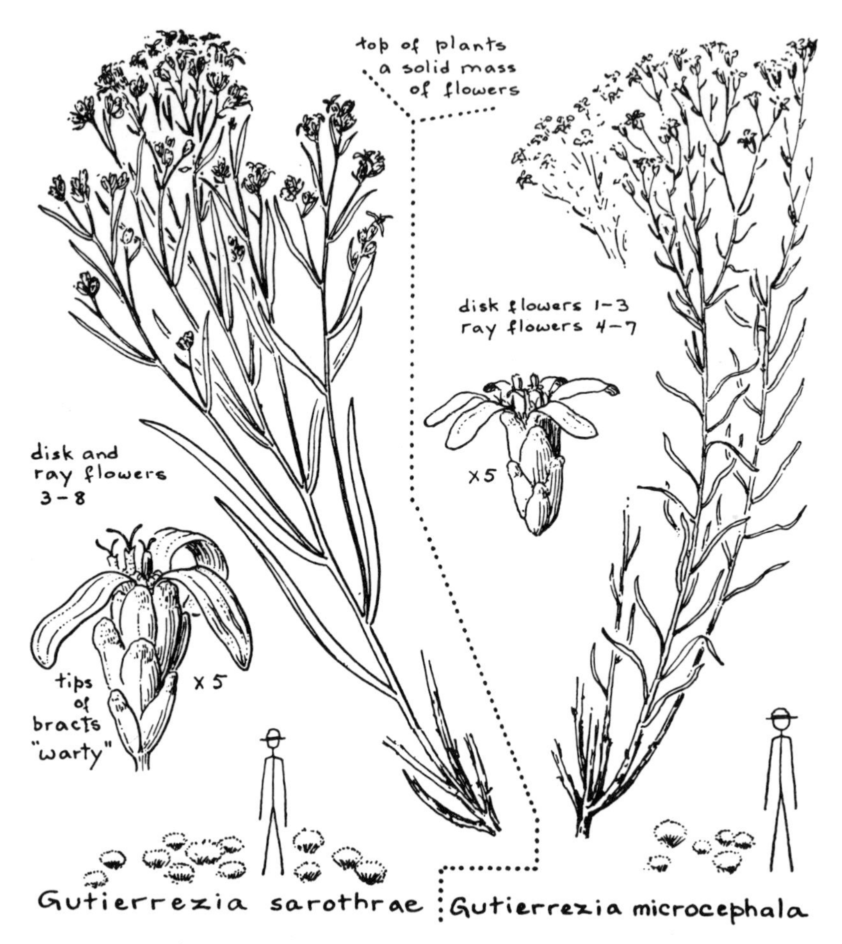

BROOM SNAKEWEED, MATCHBRUSH (Gutierrezia sarothrae)
Sunflower family *(Compositae)*. Blooms yellow, July-October.
Saskatchewan to Kansas, south to northern Mexico and Baja California, 2,800-7,000 feet.

This low, bushy, resinous plant is very common on dry stony plains and mesas all through the western United States. The yellow flowers are very small but so numerous that they are effective and often dominate the landscape in the late summer and fall.

Economically this plant is useless; it is unpalatable and somewhat poisonous to livestock, will encroach upon valuable grassland, and does not even appreciably retard soil erosion.

It is said the Navajo chew these plants and apply them to stings of bees, wasps and ants.

THREADLEAF SNAKEWEED, SMALL SNAKEWEED (Gutierrezia microcephala)
Blooms yellow, August-October.
Texas to Idaho, south to Arizona and Coahuila, 3,500-6,000 feet.

This is a similar but more delicate plant—the stems are more slender, the leaves smaller and narrower and the flowers finer. Not common, it grows at higher elevations.

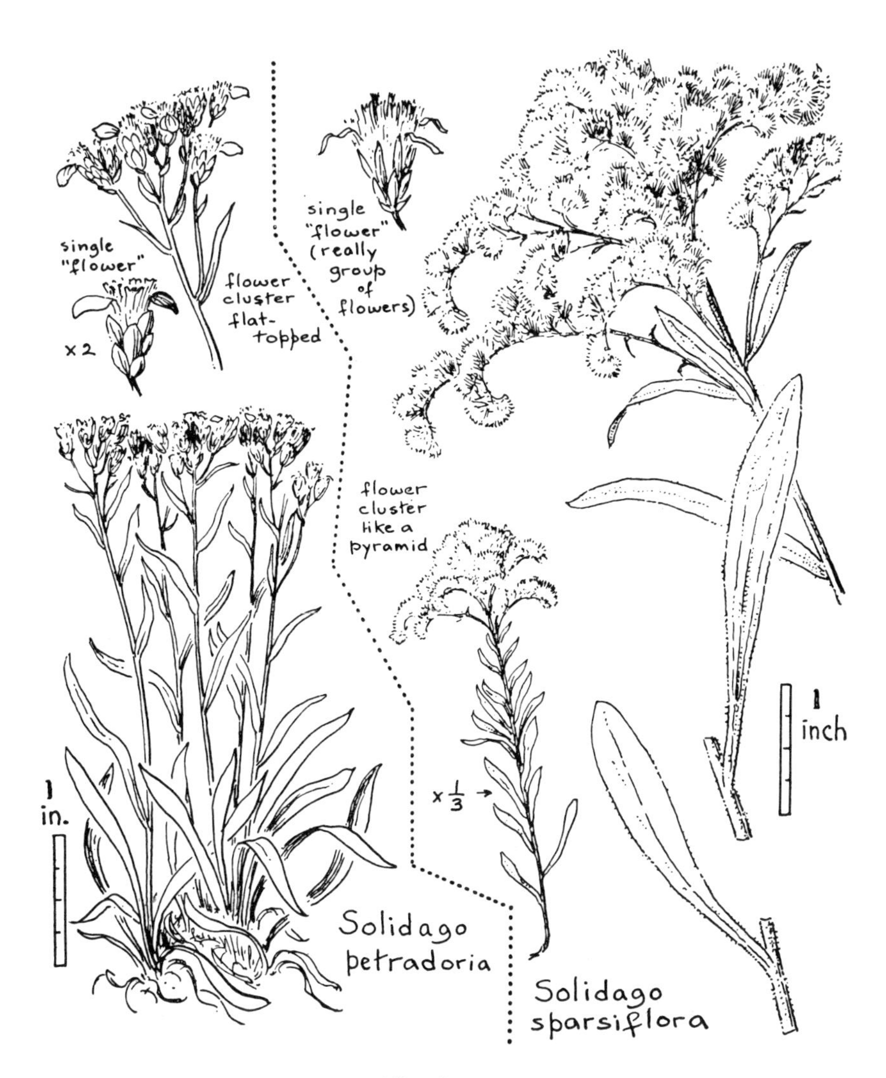

GOLDENROD (Solidago sparsiflora)

Sunflower family *(Compositae)*. Blooms yellow, June-October.

South Dakota and Wyoming to Texas, New Mexico and Arizona, 2,000-8,000 feet.

The western goldenrods are not as handsome as the eastern ones, but this species is one of the prettiest in this region. It reaches from 1 to 2 feet in height, with yellow flowers clustered on curving plume-like spikes.

The stems and three-nerved leaves are dull bluish-green, stiff and rather rough, emerging from widely spreading horizontal rootstalks.

ROCK GOLDENROD, FLAT-TOPPED GOLDENROD (Solidago petradoria)

Blooms yellow, June-August.

Wyoming to Oregon, western Texas, northern Arizona, 5,500-7,500 feet.

Superficially this small goldenrod does not resemble the one described above. The flowers are in parallel, slender, cylindrical heads; stems branch from the base; leaves are leathery.

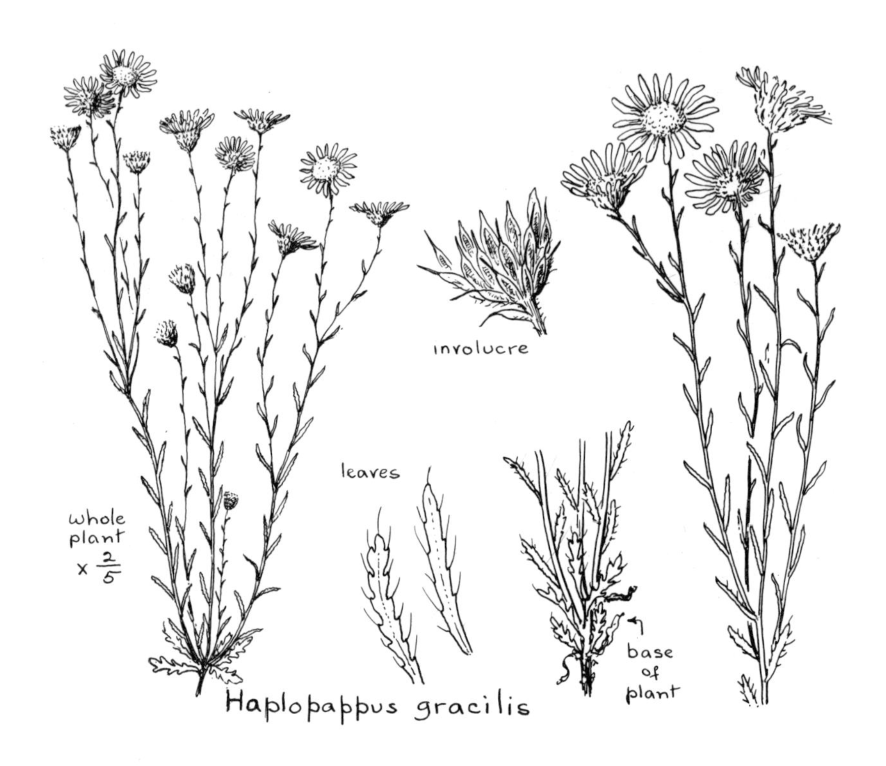

GOLDENWEED, YELLOW DAISY (Haplopappus gracilis)
Sunflower family *(Compositae)*. Blooms yellow, February-November.
Colorado to Texas, Arizona, southeastern California and Mexico, up to 6,000 feet.

Goldenweed is a slender branched annual less than 1 foot high. It associates on dry plains, mesas, and rocky slopes with the pinyon and juniper and also occurs at lower elevations in the southern desert.

Although it is a rather weedy looking plant its bright yellow flowers are pretty and bloom from early spring until late fall.

The plant is covered with small rigid hairs. The lobes of the leaves are white spiny tipped. The flower head rests in a bell-shaped involucre made up of several series of bracts that are tipped with green bristles.

The goldenweed may be seen on the south rim of Grand Canyon near Grand View Point. It also grows in Zion and Mesa Verde National Parks, Chiricahua National Monument, and in National Monuments of northern New Mexico.

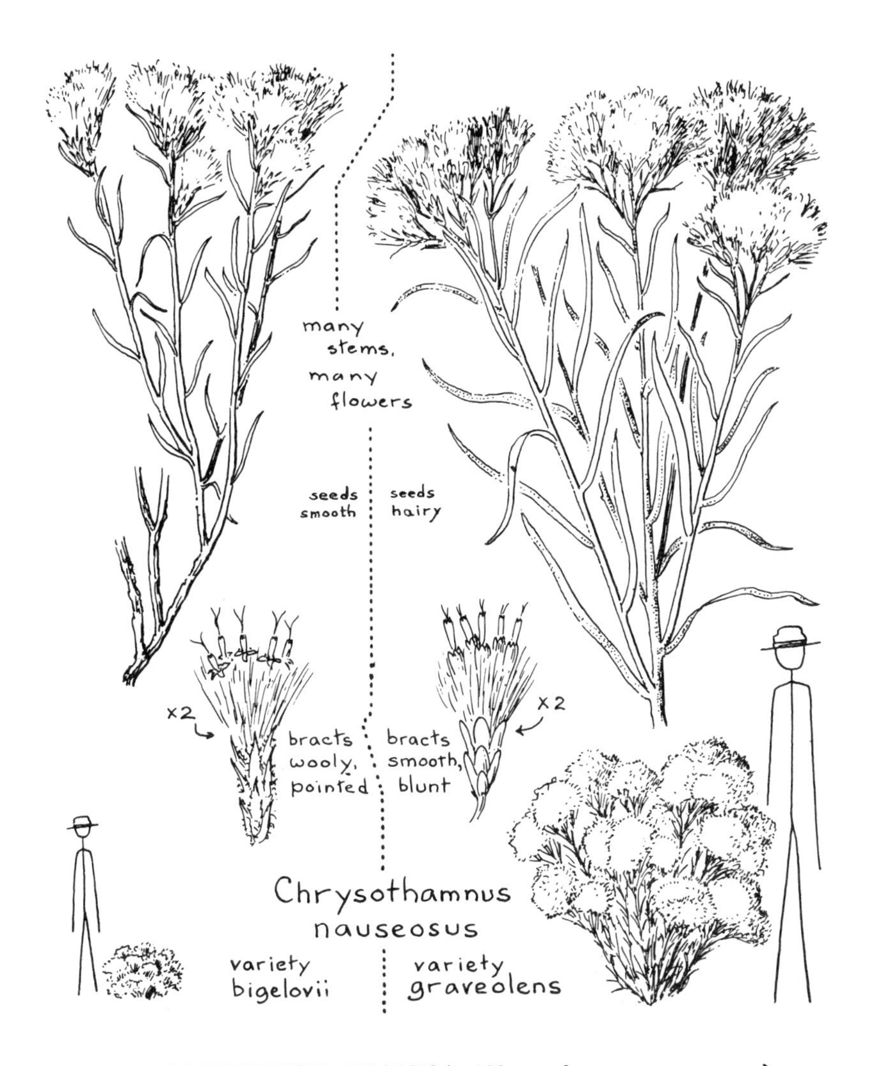

RUBBER RABBITBRUSH, CHAMISA (Chrysothamnus nauseosus)

Sunflower family *(Compositae)*. Blooms yellow, July-September.

Saskatchewan to British Columbia, south to western Texas, northern Mexico, and Baja California, 2,000-7,000 feet.

Rabbitbrush is a silvery-gray shrub crowned with bunches of feathery, golden flowers in late summer and with the white down of the seeds in the fall. It is used locally in landscaping, and is commonly seen growing around the adobe houses at Santa Fe.

There are many varieties of this widely distributed species. Two that occur at Santa Fe are illustrated here: the large one (var. *graveolens*) and the small one, (var. *bigelovii*).

The juice of this plant yields rubber but is little used commercially. A yellow dye is made from the flowers, and a green dye from the inner bark. The plant has little value as forage. The Hopi use rabbitbrush as kiva fuel, for wickerwork, as windbreaks, and for making arrows.

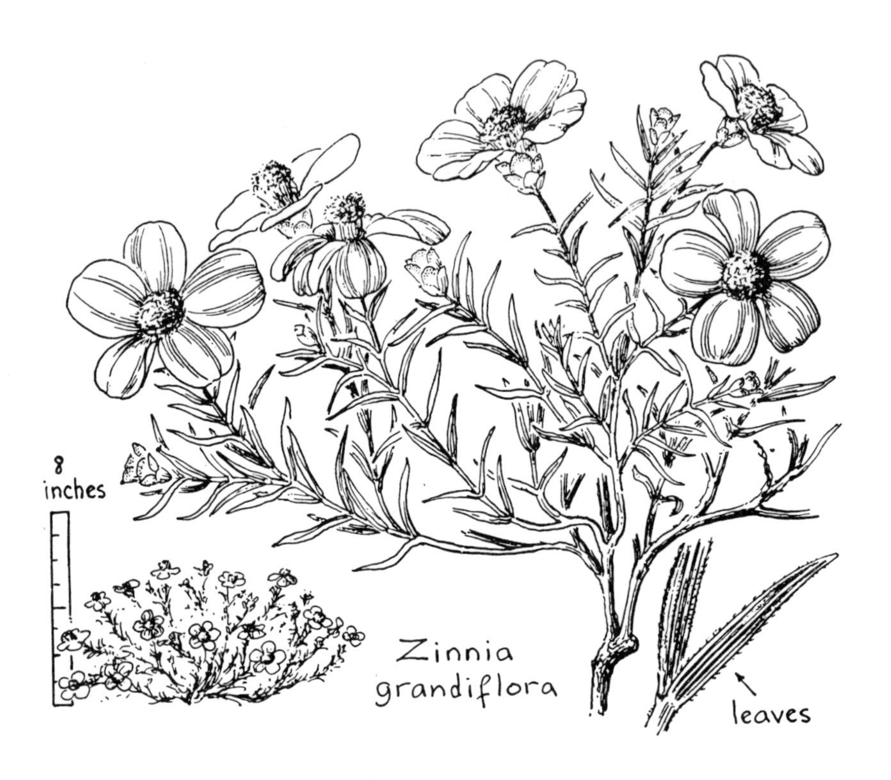

WILD ZINNIA, ROCKY MOUNTAIN ZINNIA (Zinnia grandiflora)
Sunflower family *(Compositae)*. Blooms bright yellow, June-October.
Kansas to Nevada, south to Texas, Arizona and northern Mexico, 4,000-6,500 feet.

This pretty, hardy flower of the dry mesas is sufficiently attractive to try under cultivation for ornamental borders. It is a spreading herb much branched from a woody base.

The showy yellow flower heads grow singly on the ends of the branches, and the petal-like rays are broad and about three-quarters of an inch long. Opposite, three-ribbed leaves are narrow and rather rigid.

Found in Petrified Forest National Park east of Painted Desert Inn, it is also reported from Wupatki National Monument and from White Sands National Monument far to the south. It appears in the vicinity of Santa Fe, Albuquerque, and Zuñi in northern New Mexico.

A similar flower with white rays *(Zinnia pumila)* is a native of the more southern desert.

The popular garden zinnia *(Zinnia elegans)* came from Mexico.

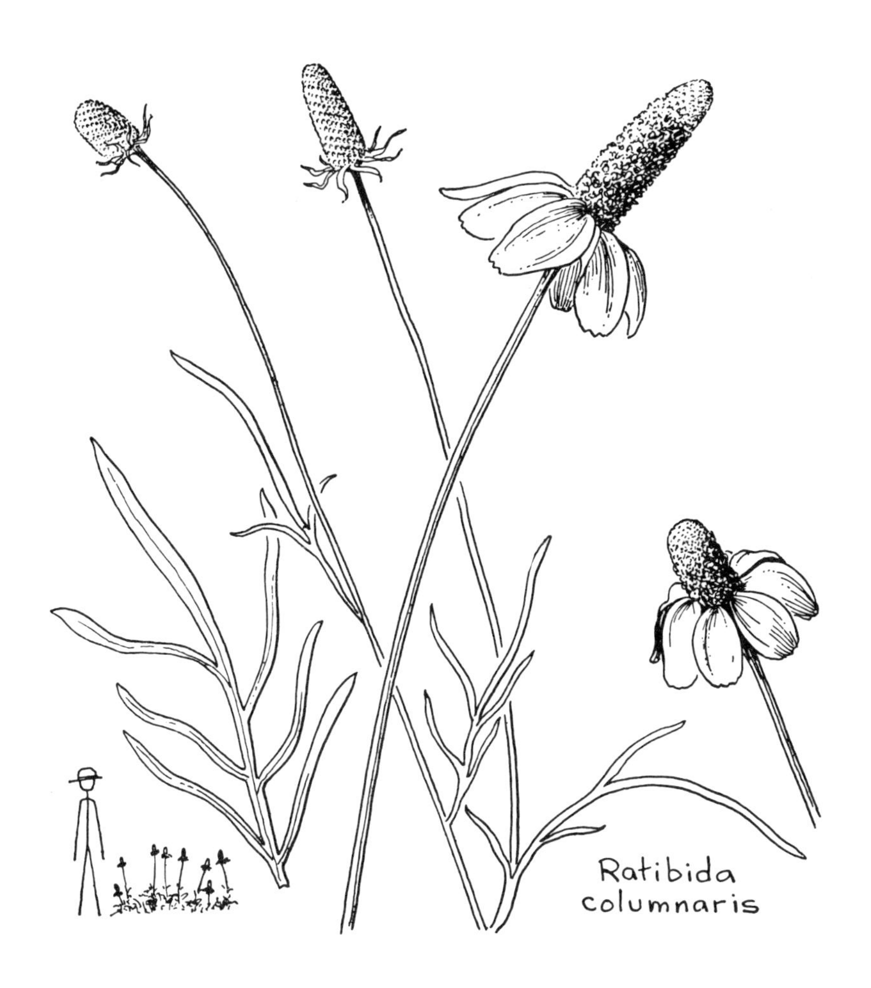

YELLOW CONEFLOWER (Ratibida columnaris)

Sunflower family *(Compositae)*. Blooms yellow, June-November.

Minnesota to British Columbia, south to Tennessee, Colorado and Arizona, 5,000-7,000 feet.

Yellow coneflower makes a striking appearance because of the fleshy, protruding center of the flower head which is surrounded by drooping petal-like rays. Plains and pine woods are its habitat in the mesa country. It may be found in the hills near Santa Fe, and on the south rim of Grand Canyon.

It is a rather slender plant. The alternate leaves are divided into long narrow lobes and the flower head grows at the top of a long swaying stalk. The tiny deep yellow flowers in the center of the cluster grow on a long cylindrical projection. Petal-like rays are usually yellow, but sometimes partly brown-purple.

The coneflower is suspected of being poisonous to cattle but is rarely eaten by them.

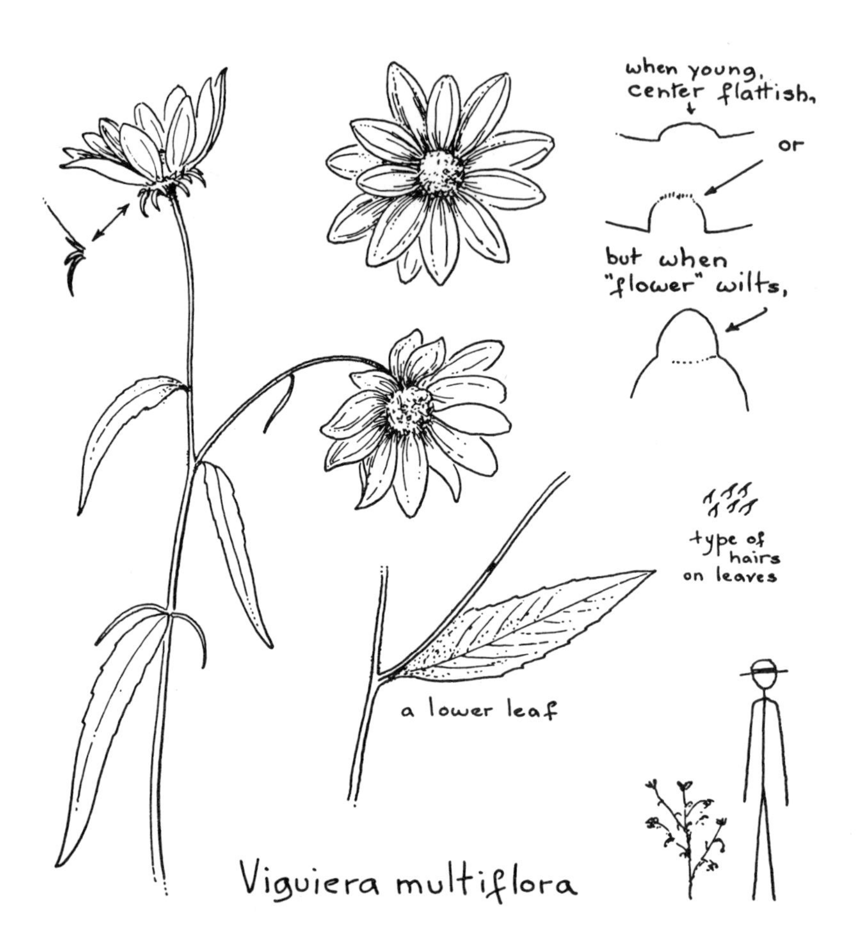

MANY-FLOWERED VIGUIERA (Viguiera multiflora)
Sunflower family *(Compositae)*. Blooms yellow, May-October.
Southwestern Montana to New Mexico, southern Arizona, Nevada and eastern California, 4,500-8,000 feet.

A slender, branching plant with dark green leaves commonly growing on dry slopes in the pinyon and ponderosa pine forests. The bright yellow petal-like rays overlap one another. The center of the flower head is darker yellow flecked with brown and is hemispherical in shape when the flowers are young, but in age becomes elevated until it is cone-shaped.

The little green leaf-like bracts of the involucre in which the flower head rests have an unusual arrangement. They are in three tiers: the lowest ones are long and turn down; the middle ones are short and turn down; the upper ones are short and turn up. The little hairs on this plant have an unusual shape, being larger at the base.

Viguiera is common on both rims of Grand Canyon. It also occurs in Mesa Verde National Park and in Walnut Canyon National Monument. A narrow-leaved variety of this species (var. *nevadensis*) is also found at Grand Canyon, and in Zion National Park.

KANSAS SUNFLOWER (Helianthus annuus)

Sunflower family *(Compositae)*. Blooms yellow with brown centers, March-October.

Saskatchewan to Texas and westward. Has become established elsewhere in the United States; 100-7,200 feet.

Kansas' state flower is the sunflower, one of the most common and conspicuous plants of roadsides and fields. It has a large handsome flower head with golden-yellow petal-like rays and a purple-brown center. Stems and dull green leaves are rough with stiff hairs.

Useful as well as handsome, it is often cultivated. It yields good honey and yellow dye can be made from the flowers. The Hopi extracted purple and black dyes from the seeds for coloring baskets and cloth and for painting their bodies in certain ceremonies.

A useful oil can be extracted from the seeds, leaves are good for fodder, and in the Plains states the sunflower is used for silage. A fiber for textile making comes from the stalks.

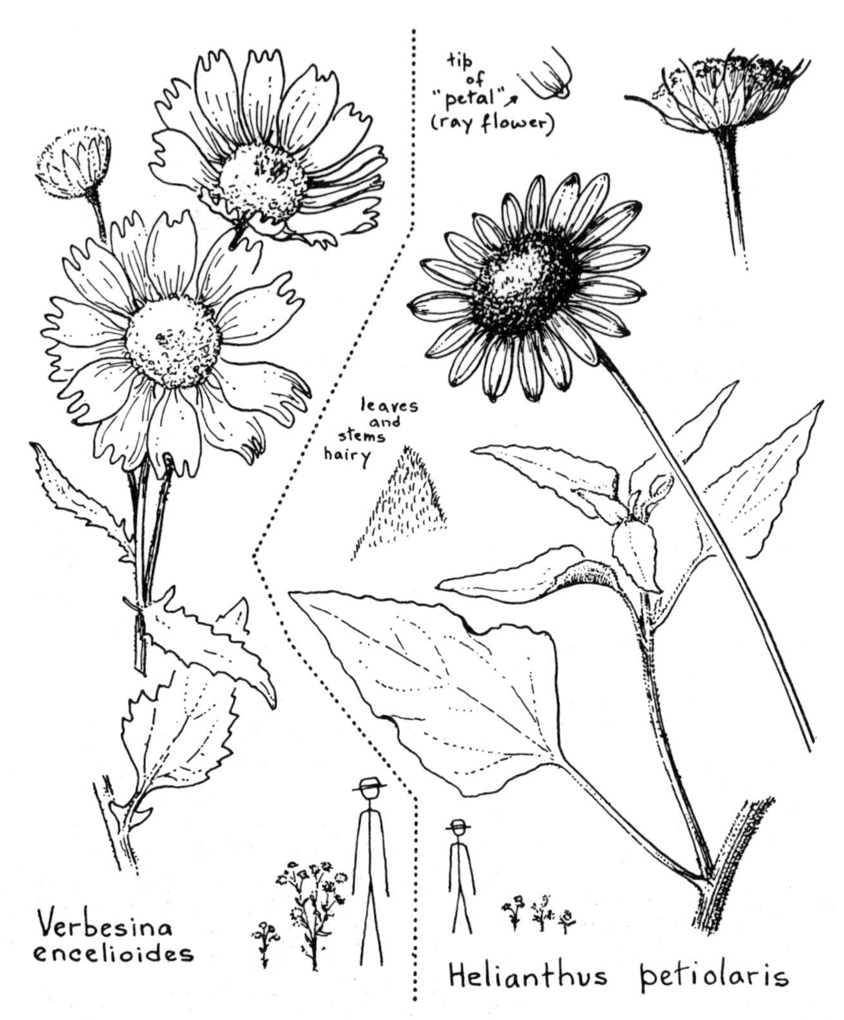

GOLDEN CROWNBEARD, GOLDWEED (Verbesina encelioides)
Sunflower family *(Composiate)*. Blooms yellow, centers flecked with brown, April-November.
Kansas to Montana, south to Texas, California and northern Mexico, up to 7,200 feet.

Golden crownbeard is a gay flower often thick in fields and waste places, making large splashes of rich yellow across the landscape almost everywhere throughout the United States.

Indians and white pioneers applied it to boils and skin diseases. The Hopi bathe in water in which the plant has been soaked to relieve spider bite.

PRAIRIE SUNFLOWER, NARROWLEAF SUNFLOWER (Helianthus petiolaris)
Same family. Blooms yellow with a purple-brown center, March-October.
Saskatchewan to Missouri and Texas, west to British Columbia and California, 1,200-7,500 feet.

One of the common sunflowers somewhat resembling the Kansas sunflower, but is a smaller, neater, more delicate plant with more compact flower heads.

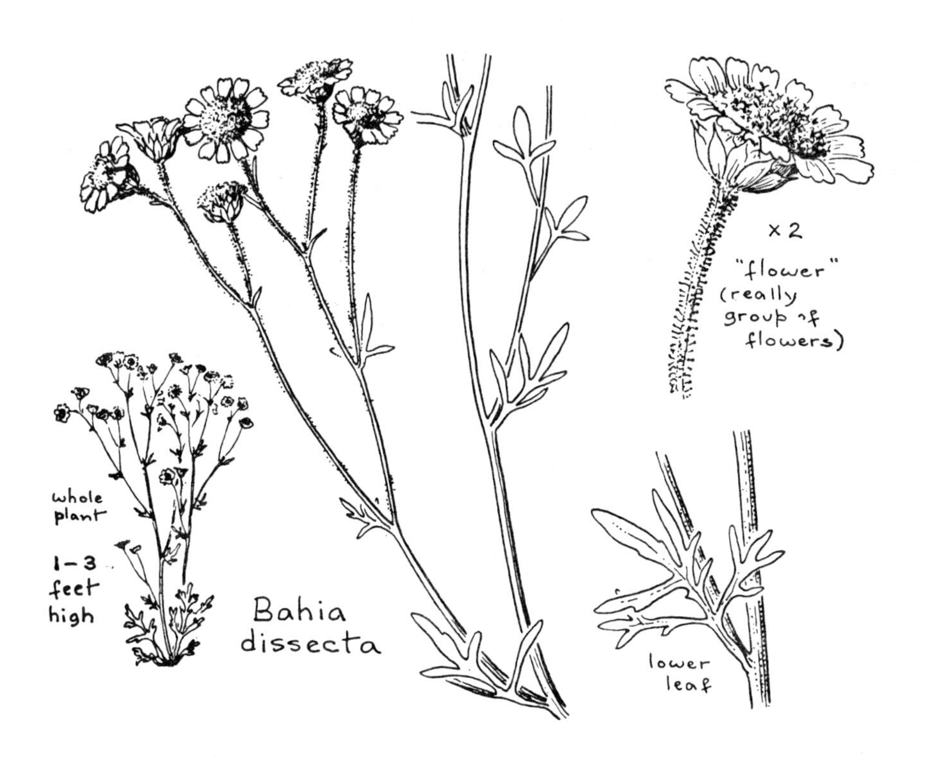

RAGLEAF BAHIA, WILD CHRYSANTHEMUM (Bahia dissecta)
Sunflower family *(Compositae)*. Blooms yellow, August-October.
Wyoming to northern Mexico and Arizona, 5,000-9,000 feet.

This golden wild chrysanthemum is common in grasslands and in open forests, growing with the pinyon and juniper trees, and also at higher elevations with the ponderosa pine and aspen. At Grand Canyon it occurs on both rims but is most abundant on the south rim. At Mesa Verde National Park it may be seen on the slope behind the museum at the head of Spruce Tree Canyon and it also is found at Walnut Canyon, Bandelier, Chiricahua and Wupatki National Monuments.

The yellow flowers are in small heads terminating the branches. They are rather numerous in open, branched clusters. The large centers of the flower heads are surrounded by petal-like rays.

The plant has stoutish branches and grows from 1-3 feet high. The alternate leaves are dark green and from one- to three-lobed. (Each lobe itself usually has three divisions.) The basal leaves are deeply lobed.

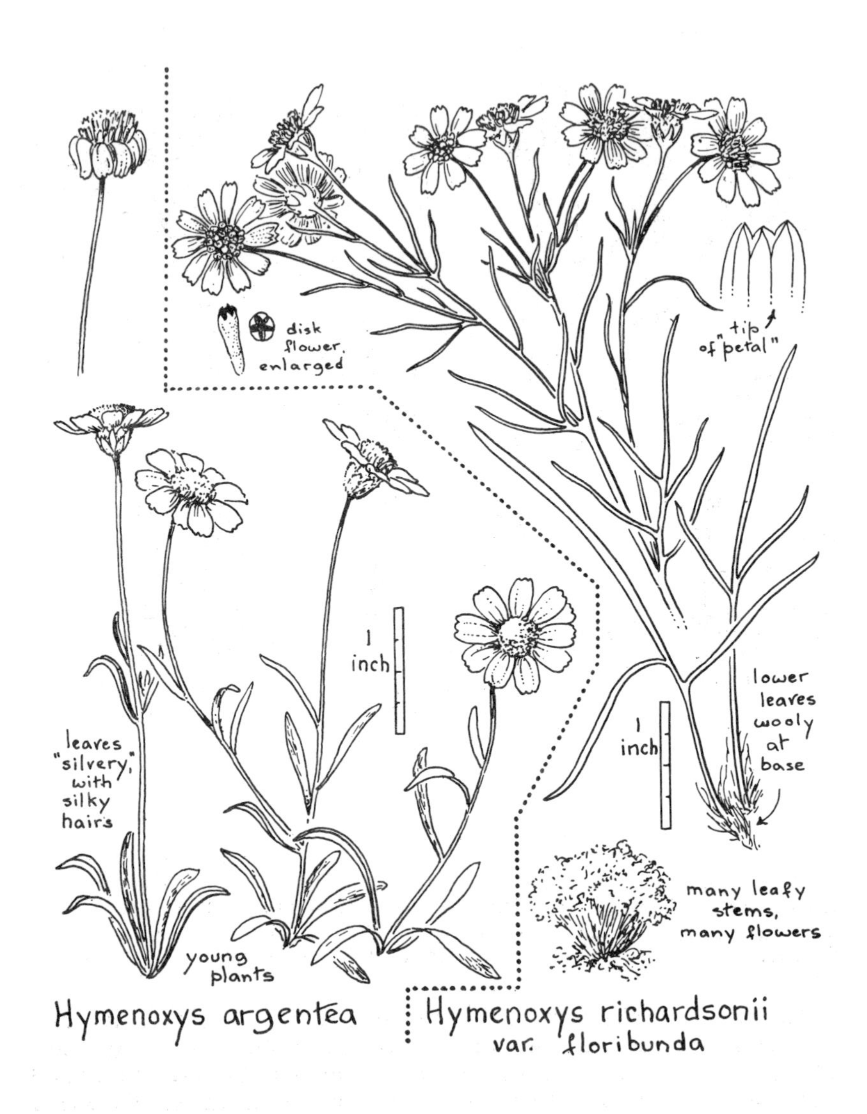

HYMENOXYS, ACTINEA, PERKY SUE (Hymenoxys argentea)
Sunflower family *(Compositae)*. Blooms yellow, April-October.
New Mexico and eastern Arizona. Pinyon-Juniper belt.

The bright yellow flowers of actinea bloom all summer and are especially profuse in the early spring and again after the summer rains. On the hills east of Santa Fe they grow quickly, tinting the landscape with gold.

The rubber plant *(Hymenoxys richardsonii* var. *floribunda)* has a smaller flower head and sometimes many more heads on a stalk, is more bushy appearing, and the leaves are cut into three lobes. It is a poisonous range plant. It too grows in the Santa Fe area and also at Grand Canyon at elevations varying from 5,200-8,000 feet.

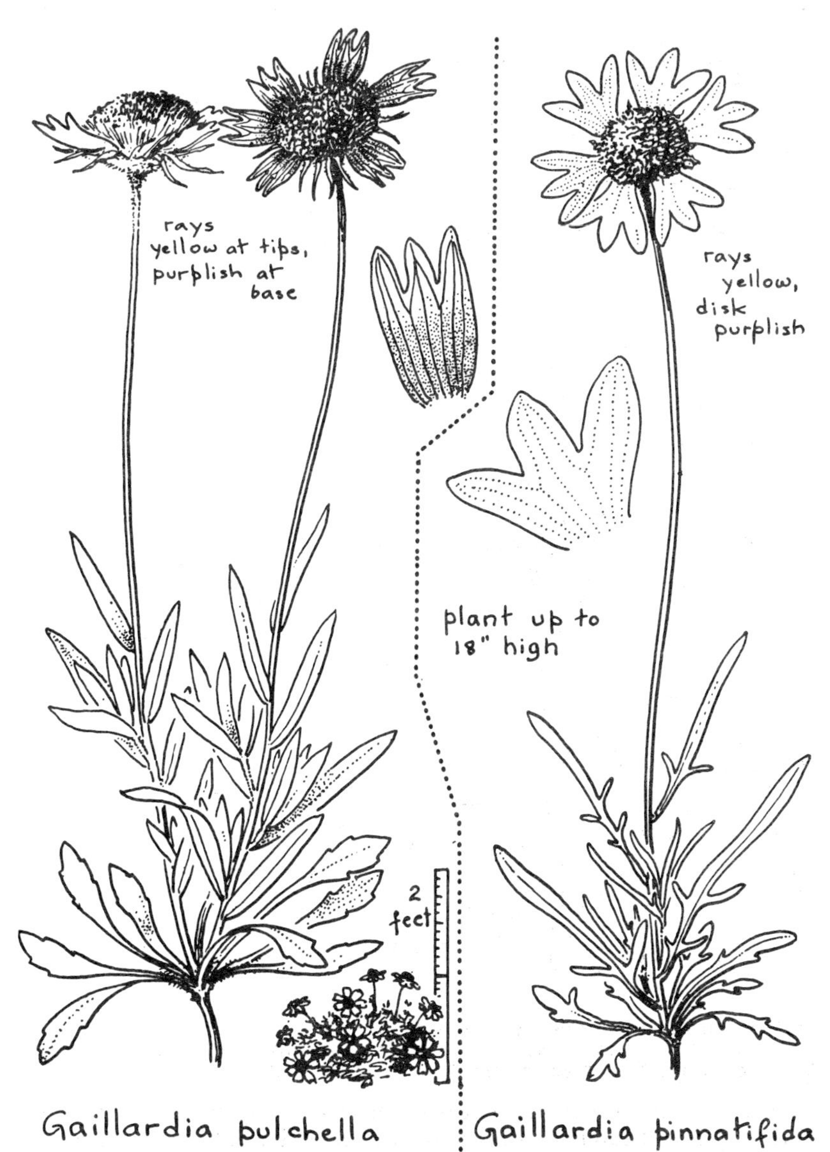

BLANKETFLOWER, FIREWHEEL (Gaillardia pulchella)
Sunflower family *(Compositae)*. Blooms yellow and purple, May-September.
Nebraska and Missouri to Louisiana, west to Colorado and southeastern Arizona, 4,000-5,000 feet.

Gaillardias are showy wildflowers and this one is particularly attractive, having three-lobed, petal-like rays which are purple at the base and yellow at the tips. Center of the flower head is purple. It is a hairy plant with leafy stems. The cultivated gaillardias are derived from this species.

PINNATE-LEAVED GAILLARDIA (Gaillardia pinnatifida)
Blooms yellow and purple, May-October.
Colorado and Utah to Texas, Arizona and New Mexico, 3,500-7,000 feet.

Has rays that are all yellow surrounding a purple velvety center; the hairy flower stalks are stiff, rough and dull green. It is a common wildflower growing on mesas, plains and in open pine forests, often on limestone.

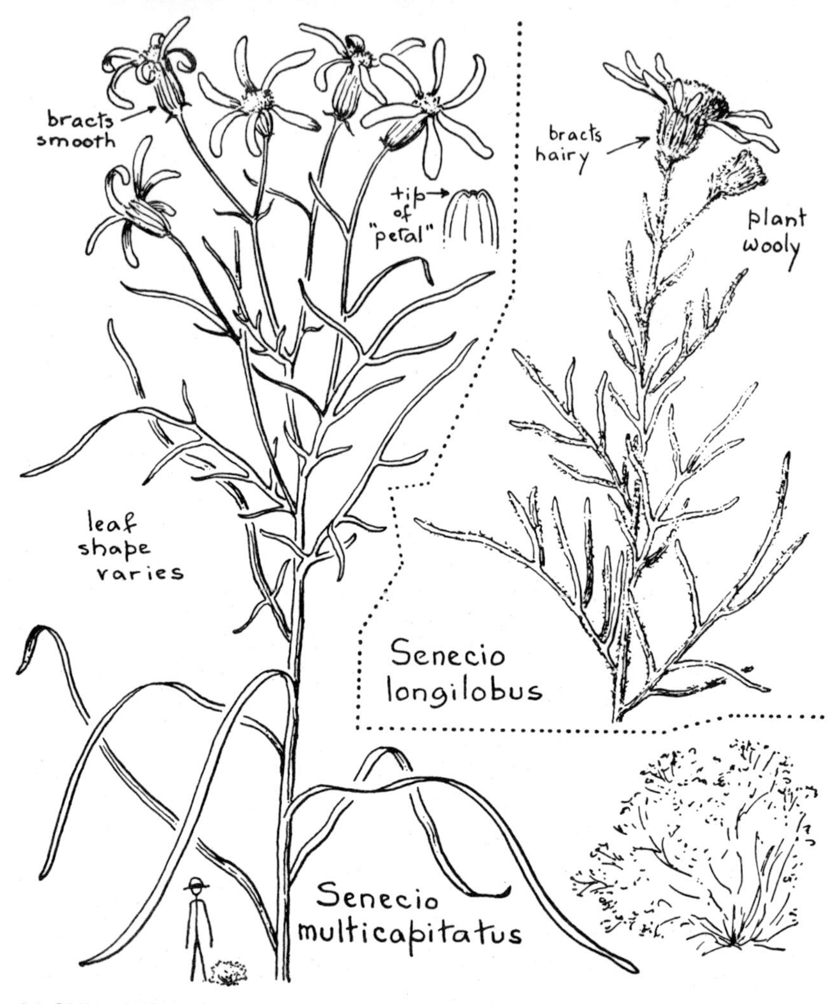

GROUNDSEL, RAGWORT (Senecio multicapitatus)
Sunflower family *(Compositae)*. Blooms yellow, May-November.
Colorado, Utah, New Mexico and Arizona, 5,000-7,000 feet.

Groundsel is a shrub-like plant covered with bright yellow flowers all through the summer. The narrow leaves give it a graceful feathery appearance.

Flower heads are rather narrow at the base; the petal-like rays are narrow and somewhat curling. The smooth, green plant is common around Santa Fe and Albuquerque. It also grows in El Morro, Navajo and Bandelier National Monuments.

THREADLEAF GROUNDSEL, FELTY GROUNDSEL (Senecio longilobus)
Blooms yellow, May-November.
Colorado and Utah, south to Texas and Mexico, 2,500-7,000 feet.

The threadleaf groundsel is very much like the one described above, but the base of the flower head is broader, and the plant is covered with matted, white, woolly hairs.

This is one of the most poisonous of the groundsels, especially to horses and cattle, producing lesions of the liver. The leaves of the new growth are most toxic. It is used extensively in the domestic medicine of the Indians. May be seen in National Monuments in northern Arizona and New Mexico.

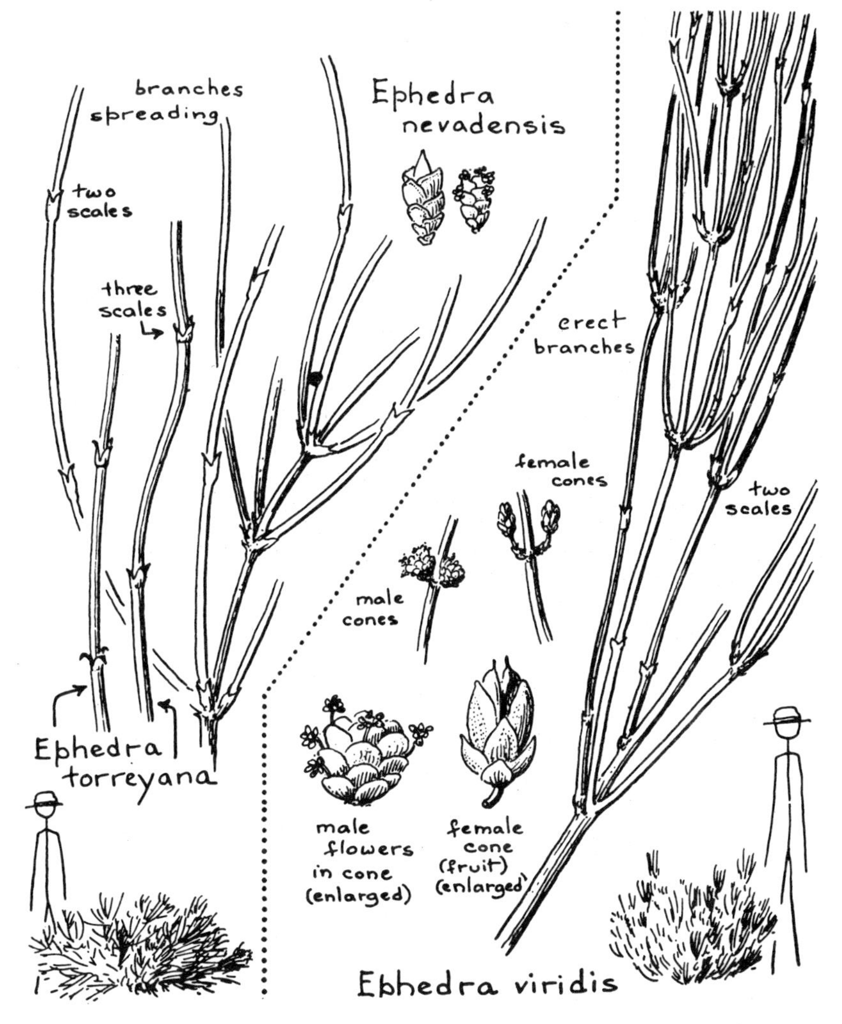

MORMON-TEA, BRIGHAM-TEA, EPHEDRA (Ephedra)
Jointfir family *(Ephedraceae)*. Flowers cone-like, green; spring.

This strange, apparently leafless plant with slender, bunched, jointed stems, is typical of the Southwest. It is a primitive type of plant with cone-like flowers; rather closely akin to the pine family. The leaves are reduced to scales. The male and female flowers grow on different plants.

Mexicans and early Mormons made tea from the stems. The plant, which contains a high percentage of tannin, was also utilized medicinally by Indians.

The three species illustrated are as follows:

(Ephedra viridis). Scales two at a joint, branches erect, bright green.
Southwestern Colorado, Utah, Nevada, Arizona and California, 3,000-7,000 feet.

(E. torreyana). Scales three at a joint, branches spreading, less rigid, olive-green.
Southwestern Colorado to Nevada, south to Texas, Arizona and Chihuahua, 4,000-6,000 feet.

(E. nevadensis). Scales two at a joint, branches spreading, bluish-green.
Oregon, Utah, Arizona and California.

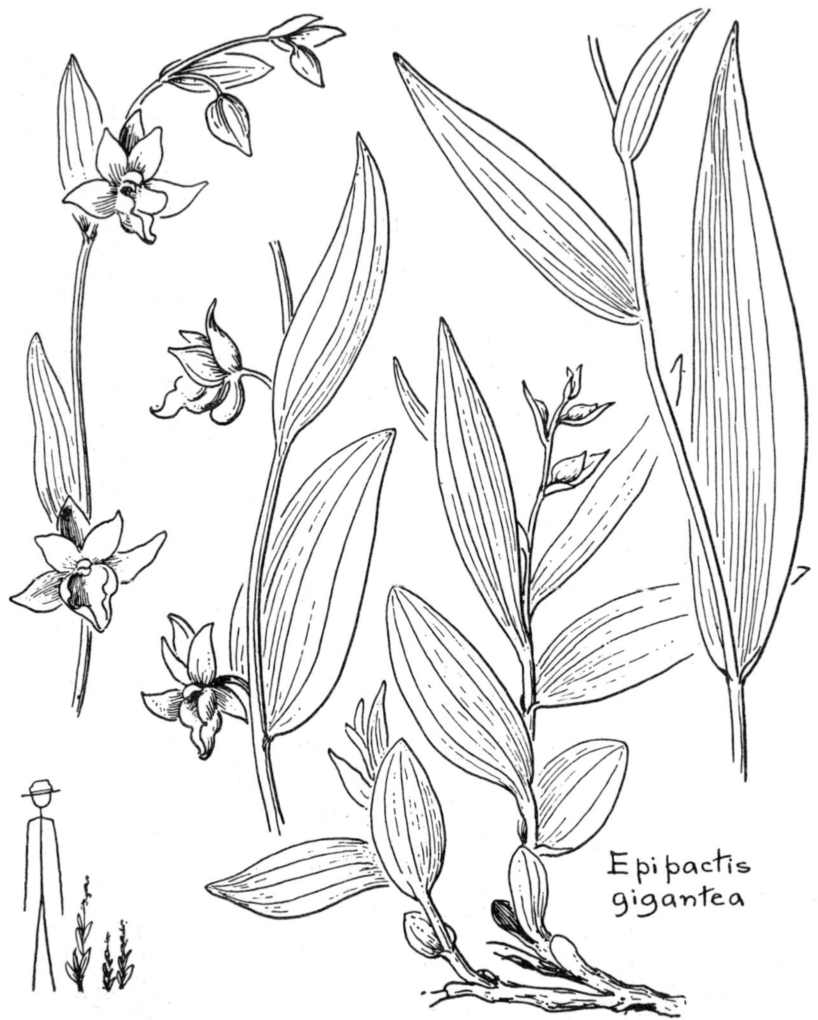

HELLEBORINE, STREAM ORCHIS (Epipactis gigantea)
Orchis family *(Orchidaceae)*. Blooms greenish, purple-veined, June-July.
Montana to British Columbia, south to western Texas, Arizona and California, 5,000-7,500 feet.

Orchids are rare in the arid Southwest, but this species may be found in rich woodland soil, in damp canyons, or under ledges of the Pinyon-Juniper Belt. It occurs near springs and in side canyons at Grand Canyon National Park and also at Keet Seel in Navajo National Monument.

It grows from 1-3 feet high from a creeping rootstalk.

There is something fascinating about the irregular flower of the orchid. This species has a greenish flower strongly veined with purple, and the lower lip of it forms a sac at the base. This lip, which swings as if from a hinge, contains the nectar and is so shaped as to accommodate the "long-tongued" insects which cross-pollinate the flower.

Many members of the orchis family are tropical and very ornamental—some are parasitic or grow on dead organic material. Extract of vanilla is made from pods of certain climbing species from tropical America.

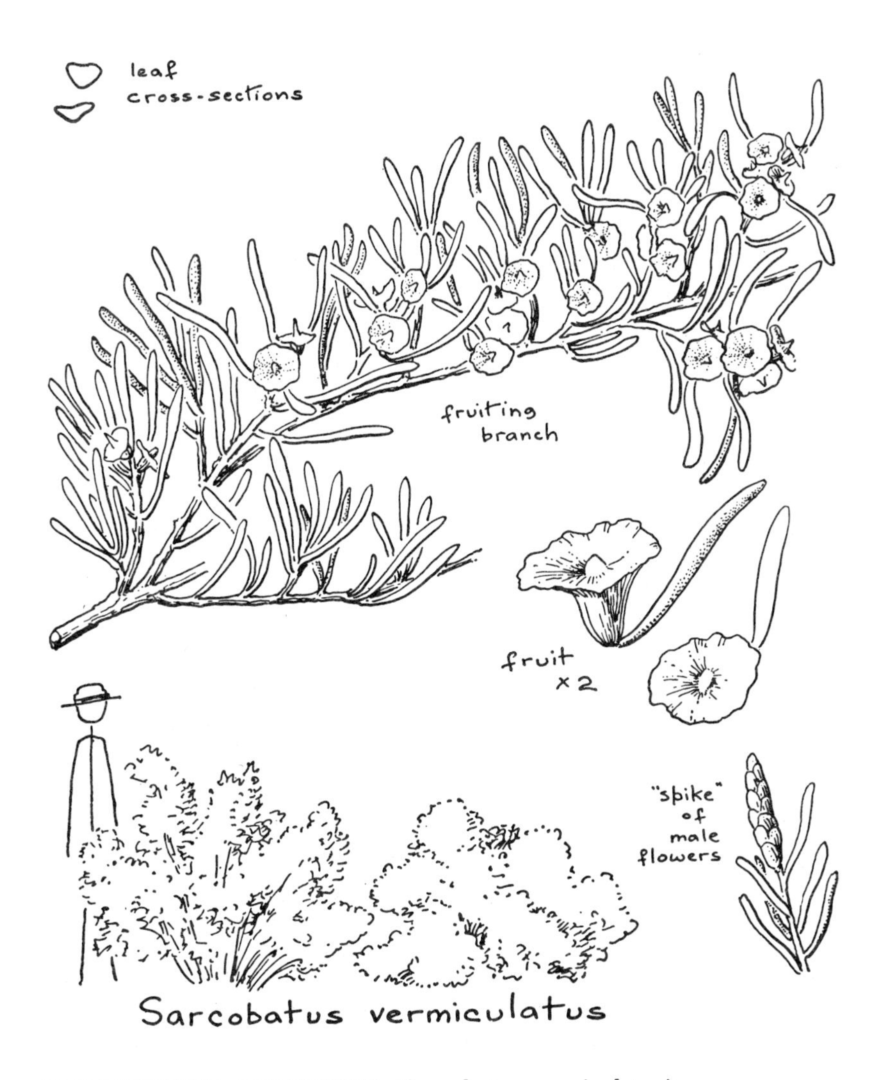

GREASEWOOD, CHICOBUSH (Sarcobatus vermiculatus)
Goosefoot family *(Chenopodiaceae)*. Flowers greenish, June-September.
North Dakota to Alberta, south to New Mexico, Arizona and California.
1,000-6,000 feet.

Greasewood grows in large colonies in wet alkaline soil. It has spreading rigid branches, with short, thornlike branchlets, and smooth white bark. The leaves are rather bright green, succulent and fleshy.

The same plant bears male and female flowers. The fruit on its stalk resembles a tiny pipe.

This is a valuable winter and spring browse plant for sheep and cattle, but can be harmful if eaten too freely. The Hopi make planting sticks from the wood, and also use it for fuel.

Greasewood may be seen in Mesa Verde and Petrified Forest National Parks and in Pipe Springs, Chaco Canyon, Aztec Ruins and Wupatki National Monuments.

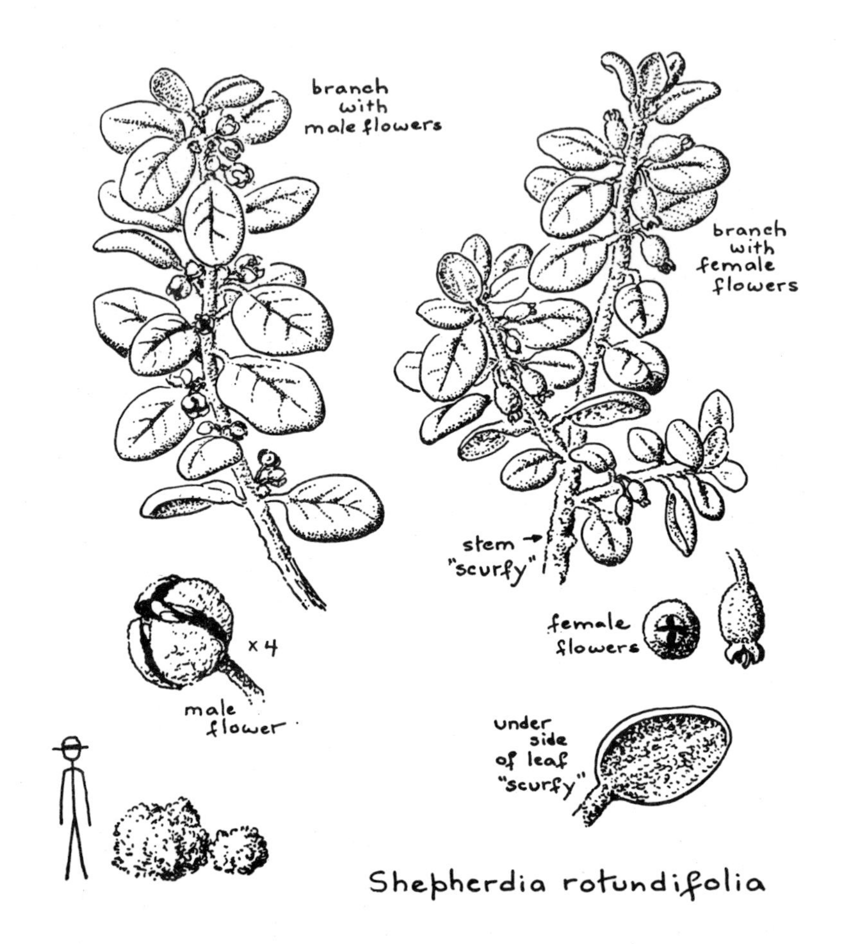

ROUNDLEAF BUFFALOBERRY, LEAD BUSH (Shepherdia rotundifolia)
Oleaster family *(Elaeagnaceae)*. Blooms gray-green, May-June.
Southern Utah and northern Arizona, 5,000-7,000 feet.

Buffaloberry is a round, compact, olive-green shrub about 3 feet high. It has a greatly limited range and apparently prefers steep rocky slopes where it is found in Zion Canyon, among rocks on both rims of Grand Canyon, and at Pipe Springs and Navajo National Monuments.

The striking and unusual features of this plant are its color and the scaly nature of the thick leaves. It looks as though it were covered with a thin coat of aluminum paint through which the green shows faintly. The rounded-oval, evergreen leaves are covered on the upper surface with silvery scales and on the under surface with yellowish-white scales and wool. Hairs on the plant are star-shaped.

The inconspicuous flowers are the same color as the rest of the plant, and grow several in the axils of the leaves. They do not have petals. The calyx is urn-shaped and has four lobes. Male and female flowers grow on different plants. The rounded berry-like fruits contain a sweet, watery, pale-yellow juice.

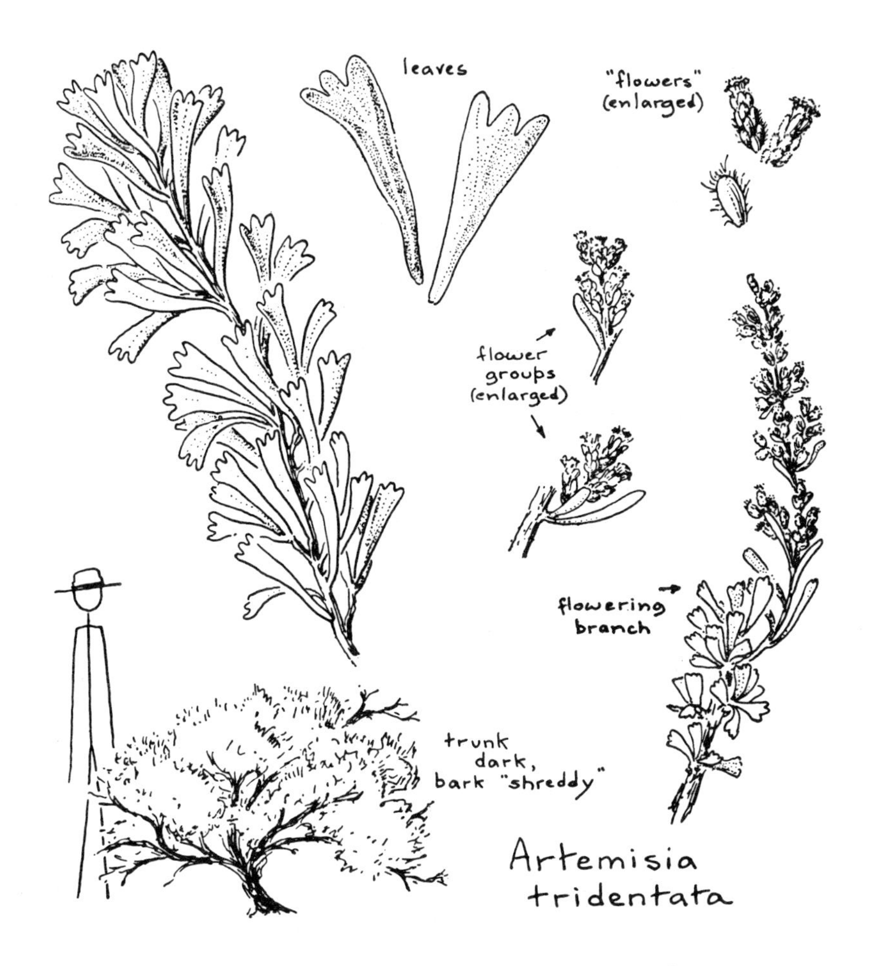

BIG SAGEBRUSH, BLACK SAGE (Artemisia tridentata)

Sunflower family *(Compositae)*. Blooms greenish or yellowish, summer and fall.

South Dakota to British Columbia, south to New Mexico, northern Arizona, and Baja California, 5,000-7,500 feet.

Sagebrush, the "purple sage" of western fiction, and the state flower of Nevada, is one of the most typical plants of the mesas and high plains of the Southwest. It is as characteristic of the northern "semidesert" or "Upper Sonoran Zone" as the creosotebush *(Larrea tridentata)* is of the southern desert. (See Natt Dodge's book "Flowers of the Southwest Deserts.")

It covers large areas in nearly pure stands where soil is deep, fertile and free from alkali. This diminutive forest gives a grayish purple tint to the landscape, because of the silvery down on the leaves and the purplish color of the shaggy bark.

The strong, aromatic odor of sagebrush is a characteristic fragrance of the West, particularly noticeable after a rain.

The size of this shrub varies greatly according to its habitat: from 1 foot, where soil is inferior, to 12 feet high, where the soil is very good.

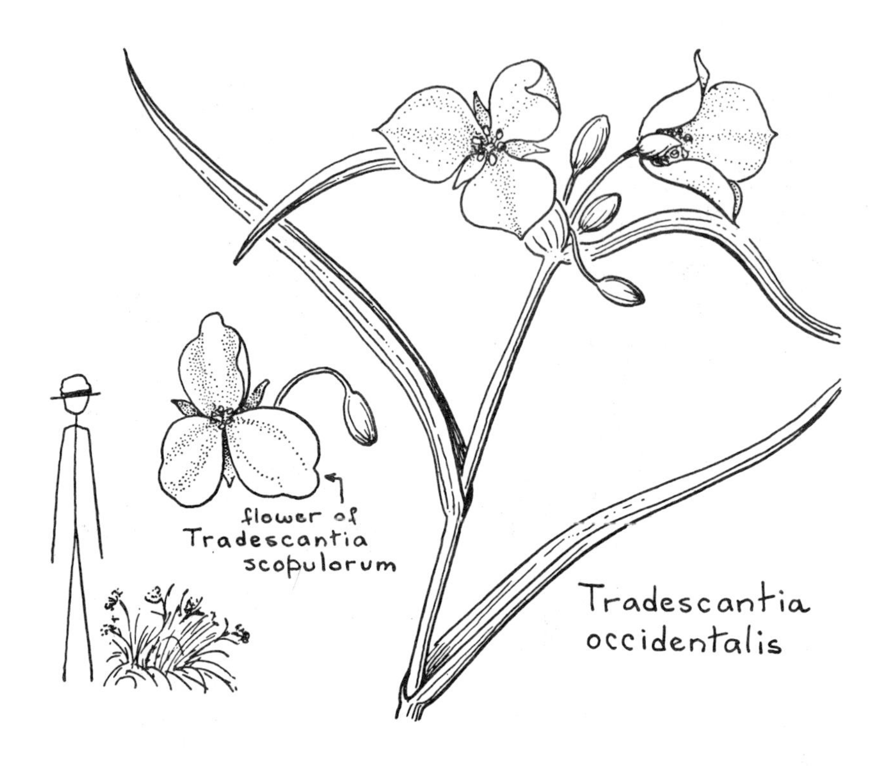

PRAIRIE SPIDERWORT (Tradescantia occidentalis)
Spiderwort family *(Commelinaceae)*. Blooms blue or reddish.
Wisconsin to Montana, Texas and Arizona, 2,500-7,000 feet.

The pretty three-petaled, blue flowers of spiderwort make an otherwise scraggly-looking plant attractive. It grows in clumps from a thick root. The green stems are stout, sparingly branched and contain a slimy sap, while the long narrow leaves are folded and clasp the stem at their bases.

It is common in Betatakin Canyon in Navajo National Monument and also is listed for the country around Aztec, New Mexico.

A closely related form (var. *scopulorum*) is without hairs. This spiderwort grows in Zion Canyon in sandy places and is particularly showy along the Narrows Trail and at the east entrance.

Still another spiderwort *(T. pinetorum)* comes out of a tuberous root. The stems of this plant do not branch, the purple flowers are small and the plant is short, under 1 foot in height.

Indians use this species as pot herbs, and they eat the root.

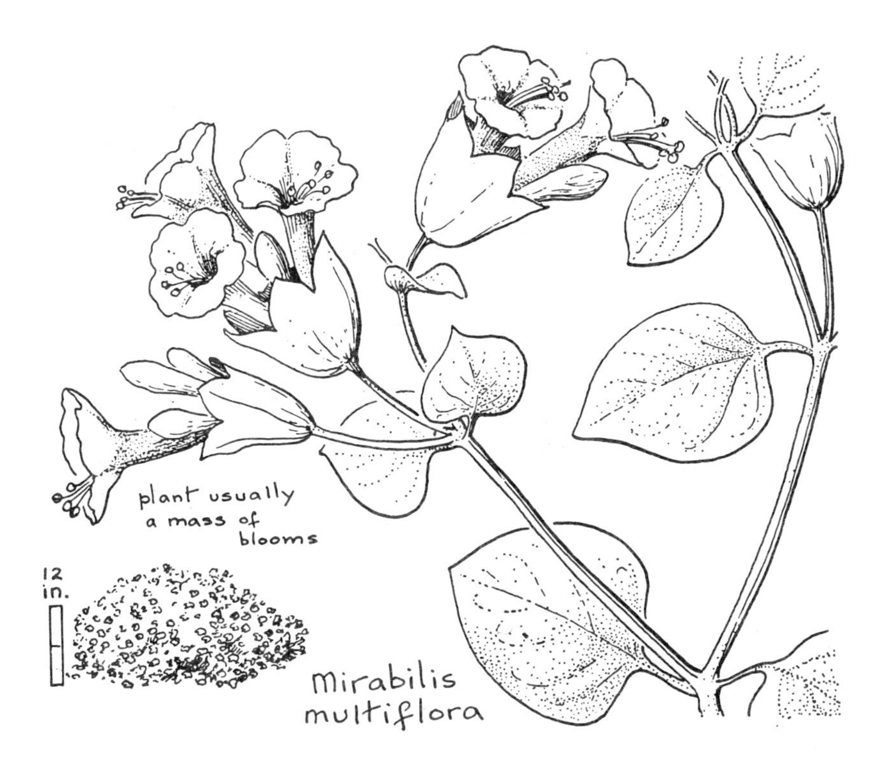

MIRABILIS, FOUR-O'CLOCK (Mirabilis multiflora)

Four-o'clock family *(Nyctaginaceae)*. Blooms magenta, April-September. Colorado and Utah to northern Mexico, 2,500-6,500 feet.

This plant closely resembles its cultivated relative, the garden four-o'clock. It grows in a rounded clump and is covered with purplish-red flowers that open in the afternoon. It is very common in the Southwest on hills and mesas and among rocks with the pinyon and juniper trees, and also at lower elevations in the southern desert.

The wild four-o'clock looks like a shrub because of the diffuse branching of its stout stems. The dark green, opposite leaves are toothless. They are somewhat heart-shaped and smooth or sometimes only slightly soft and hairy.

The showy flowers are in groups of from three to six, each group nestled prettily in a bell-shaped, green, calyx-like involucre. The flower does not have true petals, but the reddish-purple calyx is petal-like and funnel-shaped—the base of the flower tube, greenish. The five yellow stamens and purple pistil are long and protrude beyond the flower. The fruit is smooth, dark brown or black.

The powdered root of this four-o'clock is advocated as a remedy for stomach ache. Hopis eat the root to induce visions.

The plant may be seen in Grand Canyon, Mesa Verde, Petrified Forest and Zion National Parks, and in Bandelier, Chaco Canyon, El Morro and Wupatki National Monuments.

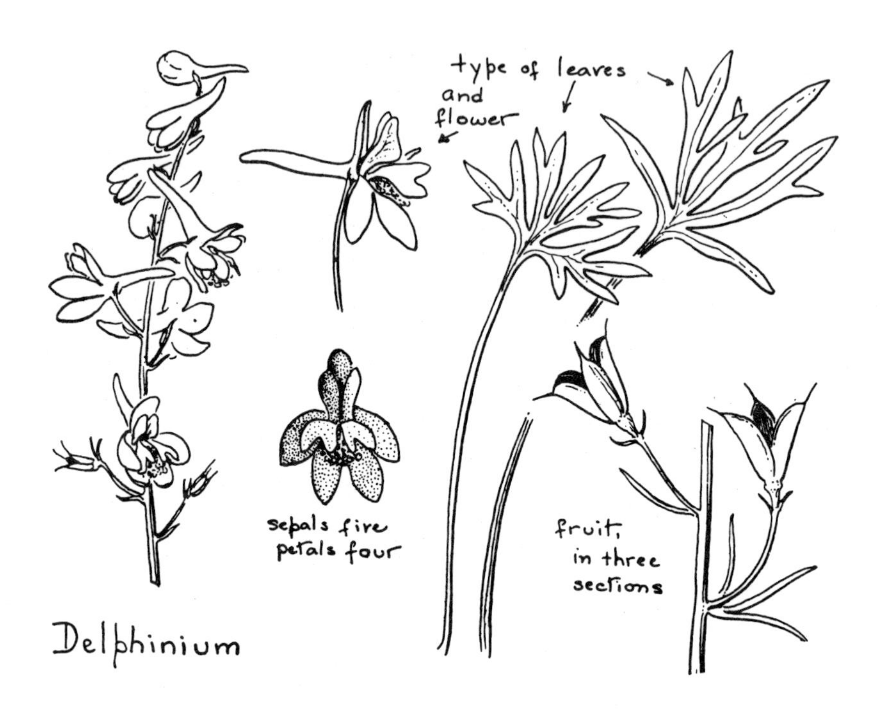

DELPHINIUM LARKSPUR (Delphinium)

Buttercup family *(Ranunculaceae)*. Blooms different shades of white, blue and purple.

There are not many species of delphinium growing in the Pinyon-Juniper Belt, but it is a thrill to find these beautiful flowers in lovely shades of blue among the gray semi-desert plants.

The drawing shows the typical flowers, leaves, and fruit of the larkspurs.

The low-growing larkspurs in the sand bloom in the spring and early summer and cluster in colonies. Those that prefer moist places are tall, and bloom in the summer and autumn. They occur in close stands in mountain meadows or partially shaded ravines.

The larkspurs contain delphinine and other poisonous alkaloids. Some species are deadly poison to cattle, but sheep and horses are not as susceptible to this poison.

A delphinium with large royal blue flowers *(Delphinium scaposum)*, growing in the mesa country, is about 2 feet high. It is used by the Hopi who call it "tcoro'si." It is said that they grind the flowers with corn to make a blue meal, "blue pollen," for the flute altar. They also use it as an emetic in one of their ceremonies.

This species of delphinium has been reported from Grand Canyon and Petrified Forest National Parks and in Chaco Canyon, Aztec Ruins, Walnut Canyon, and Navajo National Monuments.

Another species of this region *(Delphinium menziesi)* has very dark purplish-blue flowers. A low growing plant, under 1 foot high, it is listed at Grand Canyon, Zion and Mesa Verde National Parks and in Navajo and Canyon de Chelly National Monuments.

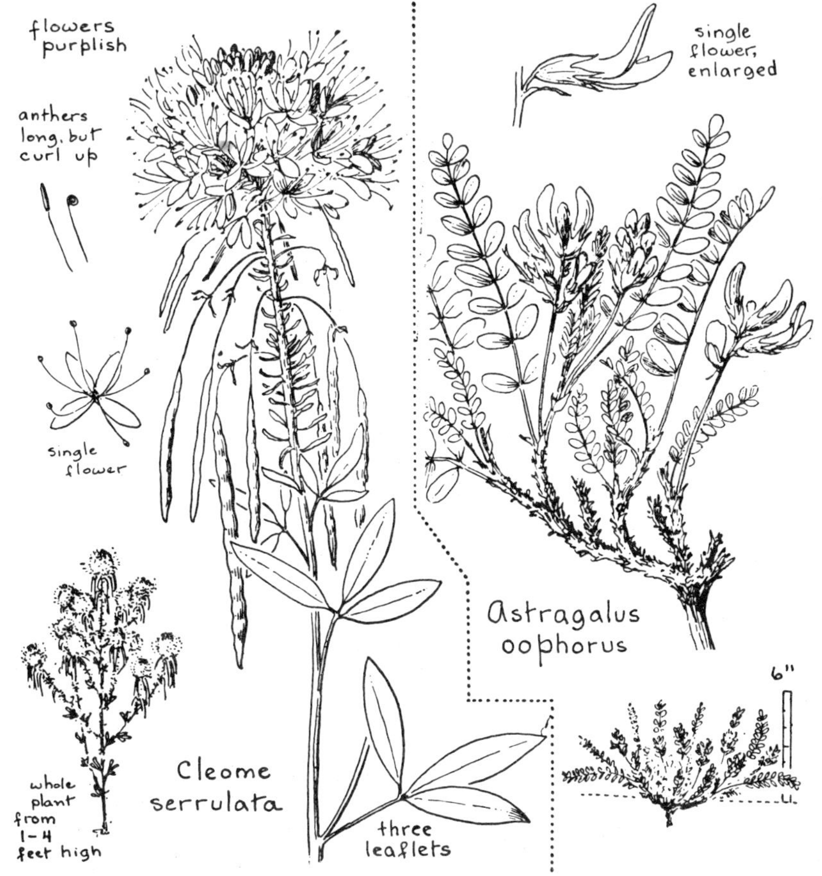

ROCKY MOUNTAIN BEEPLANT, SKUNKWEED (Cleome serrulata)
Caper family *(Capparidaceae)*. Blooms purplish-pink, June-September.
Saskatchewan to Kansas, Arizona and Oregon, 4,500-7,000 feet.

A tall, quite beautiful plant common along the roadside. The large cluster of orchid-colored flowers has a feathery appearance because of the long stamens that are tipped with green anthers. The buds at the top of the flower head are purple and green pods hang down from their slender stems below the flowers, which provide large quantities of nectar and are an important source of honey.

The leaves have three soft, smooth, bluish-green leaflets that have an unpleasant odor when crushed, hence the name skunkweed.

The yellow beeplant is shown on page 59.

ASTRAGALUS, LOCOWEED, MILKVETCH (Astragalus oophorus)
Pea family *(Leguminosae)*. Blooms white and purple, May.
Colorado and northern Arizona, 5,000-7,000 feet.

This species of milkvetch is showy when it is in fruit because of its large, bladder-like, thin-walled pods that are heavily mottled with reddish-brown. They are about 1½ inches long, and the seeds rattle in the dry ones; hence the name rattleweed. This is the common milk-vetch on the south rim of Grand Canyon, reaching about 6 inches high.

Some species of this group cause loco disease in livestock, especially horses, and are therefore called locoweed.

A closely related species with cream-colored flowers is shown on page 28.

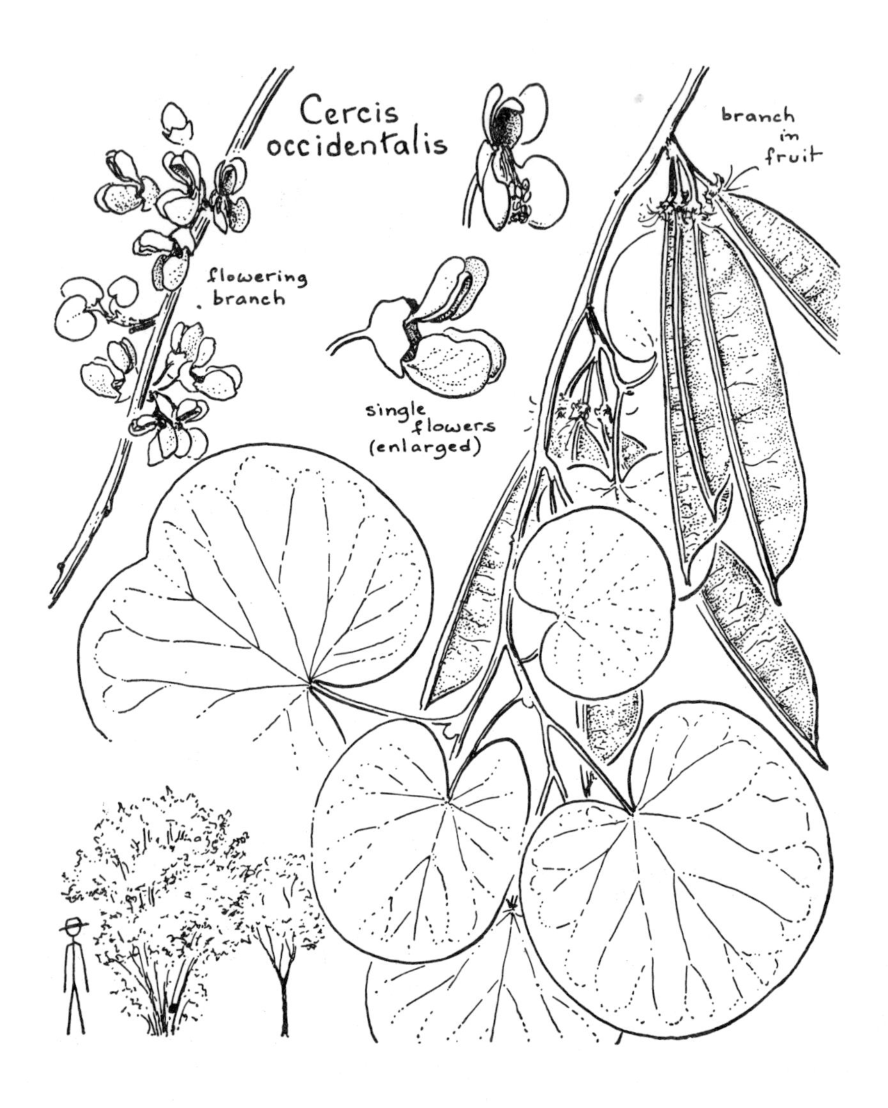

CALIFORNIA REDBUD, JUDAS-TREE (Cercis occidentalis)
Pea family *(Leguminosae)*. Blooms purplish-red, March-April.
Southern Arizona and California, about 4,000 feet.

This beautiful little tree is rare, but is conspicuous in the early spring in Grand Canyon along Bright Angel and Hermit trails. It grows about 12 feet high and has smooth, gray bark.

The purplish-red flowers occur in scattered groups on the old wood before the leaves appear, which are large and have an unusual shape. Seed pods are large, flat, thin-walled and contain several seeds.

Some species of redbud are grown as ornamentals. The eastern redbud *(Cercis canadensis)* has astringent bark which has been prescribed as a remedy for diarrhea and dysentery.

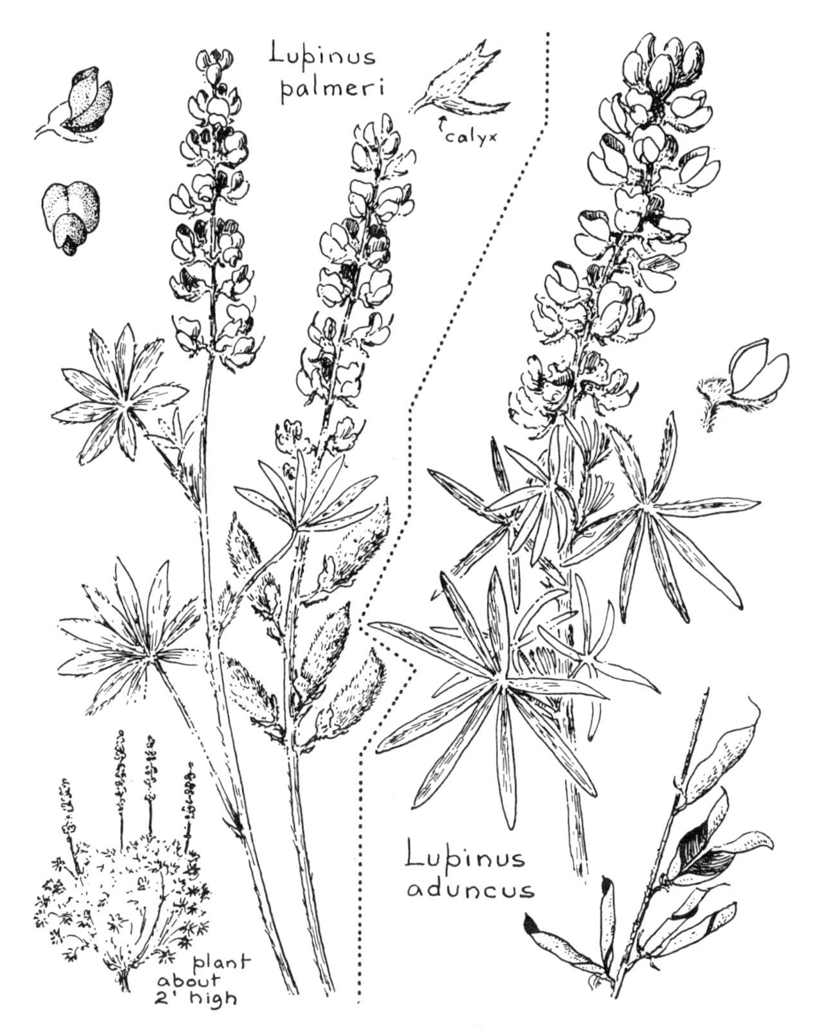

LUPINE BLUEBONNET (Lupinus palmeri)
Pea family *(Leguminosae)*. Blooms normally violet, May-October.
New Mexico and Arizona, 4,000-8,000 feet.

Lupines are among our most beautiful wild flowers, often forming large patches or fields, coloring the landscape with shades of violet and blue. The silvery-green plants have attractive leaves, the leaflets radiating from a common center.

Various species occur in nearly all of the National Park areas of this region. This species is distinguished by its spurless calyx, spreading hairs and violet flowers. It is common in Grand Canyon and Zion National Parks.

BLUE LUPINE (Lupinus aduncus)
Blooms pale purple or blue, June-September.
Wyoming and Utah to New Mexico and northern Arizona, 5,500-9,000 feet.

This is the common species in the Santa Fe region and grows in Aztec Ruins and Bandelier National Monuments. It is characterized by its short-spurred calyx, silky, silvery foliage and young stems, and its light-blue flowers.

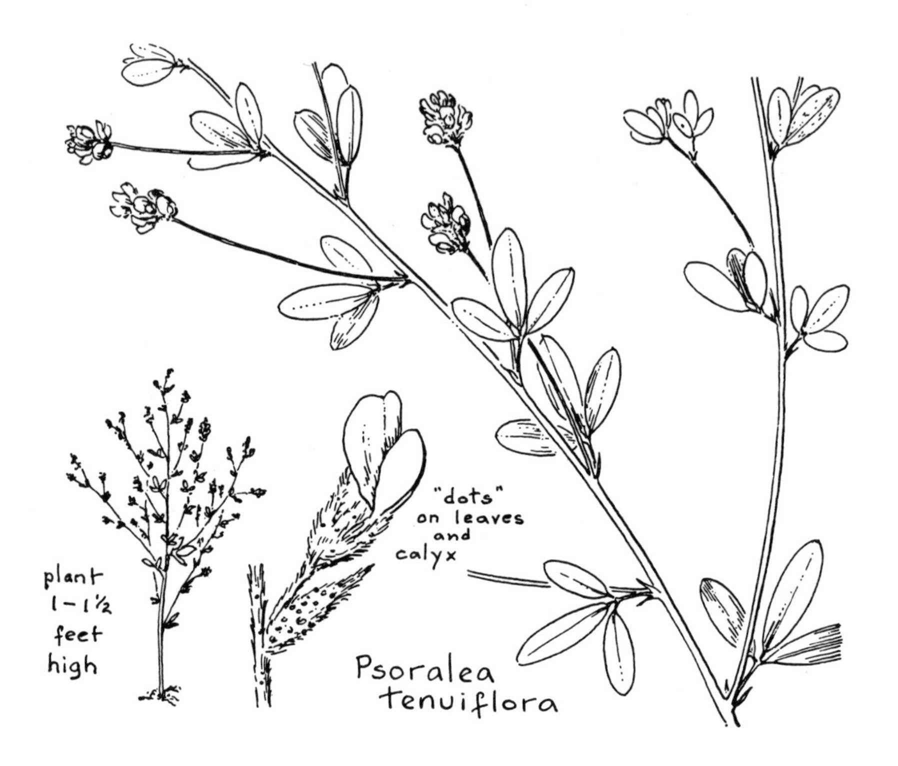

SCURF-PEA, INDIAN BREADROOT (Psoralea tenuiflora)
Pea family *(Leguminosae)*. Blooms violet, May-August.
North Dakota and Montana to Arizona and northern Mexico, 4,000-7,000 feet.

A rather common member of the pea family, liking dry slopes and plains and, often, pine woods. Although the flowers are small they are quite striking because of their deep violet color.

The plant has leafy branched stems about 1-1½ feet high. The green leaves are divided into three leaflets dotted with glands that are visible when held against the light.

The small, pea-shaped flowers form clusters on the ends of long, slender, leafless branchlets. Only one seed is contained in each small speckled pod.

It grows on both rims of Grand Canyon and in Bandelier National Monument. A white flowered species *(Psoralea lanceolata)* occurs in Walnut Canyon and El Morro National Monuments.

The tuberous roots of the scurf-peas, called breadroot, were used for food by Indians and early white settlers. This species is reported to be poisonous to horses and cattle.

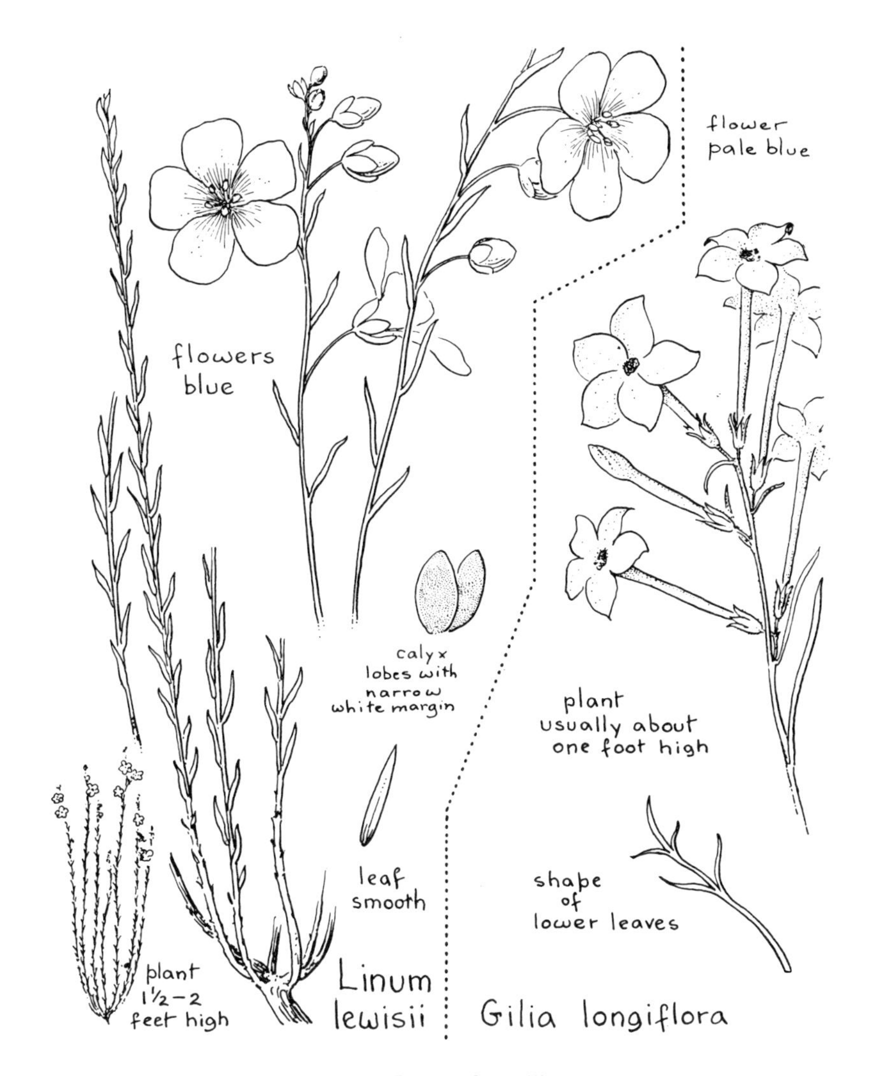

BLUE FLAX, PRAIRIE FLAX (Linum lewisii)

Flax family *(Linaceae)*. Blooms sky-blue, sometimes nearly white, March-September.

Saskatchewan and Alaska to northern Mexico, 3,500-9,500 feet.

This showy plant with its sky-blue flowers and bluish-green leaves commonly inhabits open mesas and pine woods. It is conspicuous at Grand Canyon, especially along roadsides on the north rim, and at Bryce Canyon. In Zion it may be seen above the tunnel and in cool canyons as at Emerald Pool. It also occurs at Mesa Verde National Park and at Aztec Ruins and Walnut Canyon National Monuments.

The yellow flax is shown on page 66.

BLUE GILIA (Gilia longiflora)

Phlox family *(Polemoniaceae)*. Blooms pale blue to nearly white, April-October.

Nebraska to Utah, Texas and Arizona, 1,200-6,500 feet.

Conspicuous and beautiful.

Pink and red gilias are described on page 49.

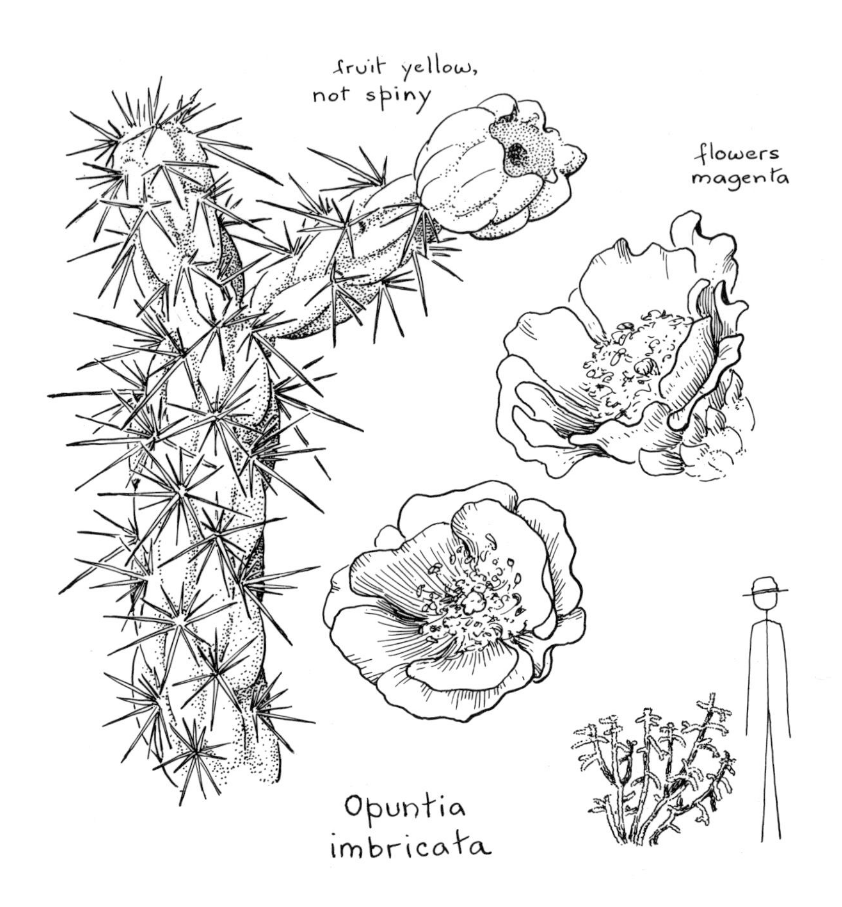

CANE CACTUS, CANDELABRUM CACTUS (Opuntia imbricata)
Cactus family *(Cactaceae)*. Blooms magenta or purple, June-July.
Colorado to Arizona and western Texas, south into Mexico, 4,500-7,500 feet.

Cane cactus is beautiful with magenta flowers that cover the ends of the branches in profusion in early summer. It is abundant especially in northern New Mexico, growing with the pinyon and juniper trees. It is common in the hills around Santa Fe. There is an especially large stand of cholla between Bernalillo and Aztec.

The spines are reddish-brown, loosely covered with a sheath that is shining white at the base, and brown-tipped; fruit is dry and yellow when mature.

The meshed wood of the stems is used to some extent for making canes, picture frames and various curios.

In their Wind Chant, the Navajo picture this cactus, which they call "biting cactus." They put the "Tall Cactus People," whom they say cure skin diseases, in the land of mirage and rocks.

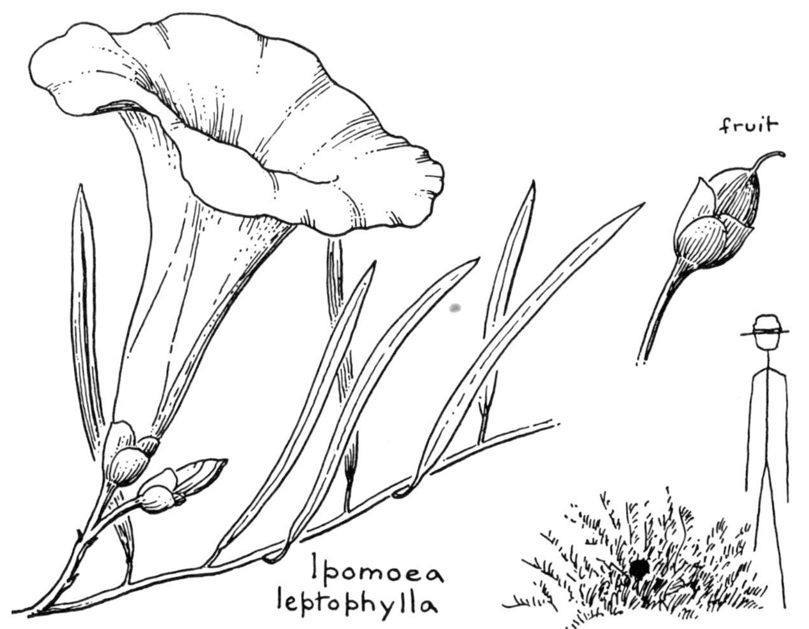

BUSH MORNING-GLORY
(Ipomoea leptophylla)
Convolvulus family
(Convolvulaceae)
Montana and South Dakota to Texas and northeastern New Mexico, 4,500-7,000 feet.

Blooms magenta. Stems, erect, forming dense clumps. Well adapted to withstand extreme drought because of an enormous root weighing up to 100 pounds and extending 4 feet into the subsoil.

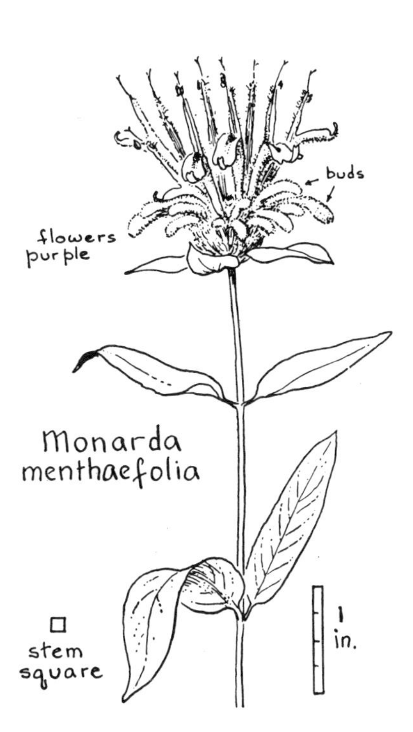

BEEBALM, HORSEMINT, BERGAMOT
(Monarda menthaefolia)
Mint family *(Labiatae)*. Blooms purple, summer.
Canada to New Mexico and Arizona, 5,000-8,000 feet.

Horsemint of this species has long purple flowers grouped in a head at the top of the stem, which is square and tinged with red or purple. It has a strong characteristic mint odor and occurs in moist places of the mesa country.

A closely related species with white flower is shown on page 32.

In the mint family are many useful plants such as sage, lavender, thyme, mint and rosemary. This plant is of some value as a forage plant and yields good honey. It is sometimes cultivated by the Hopi, who use it as a pot herb, drying the plants for use in winter. Some species are utilized in domestic medicine.

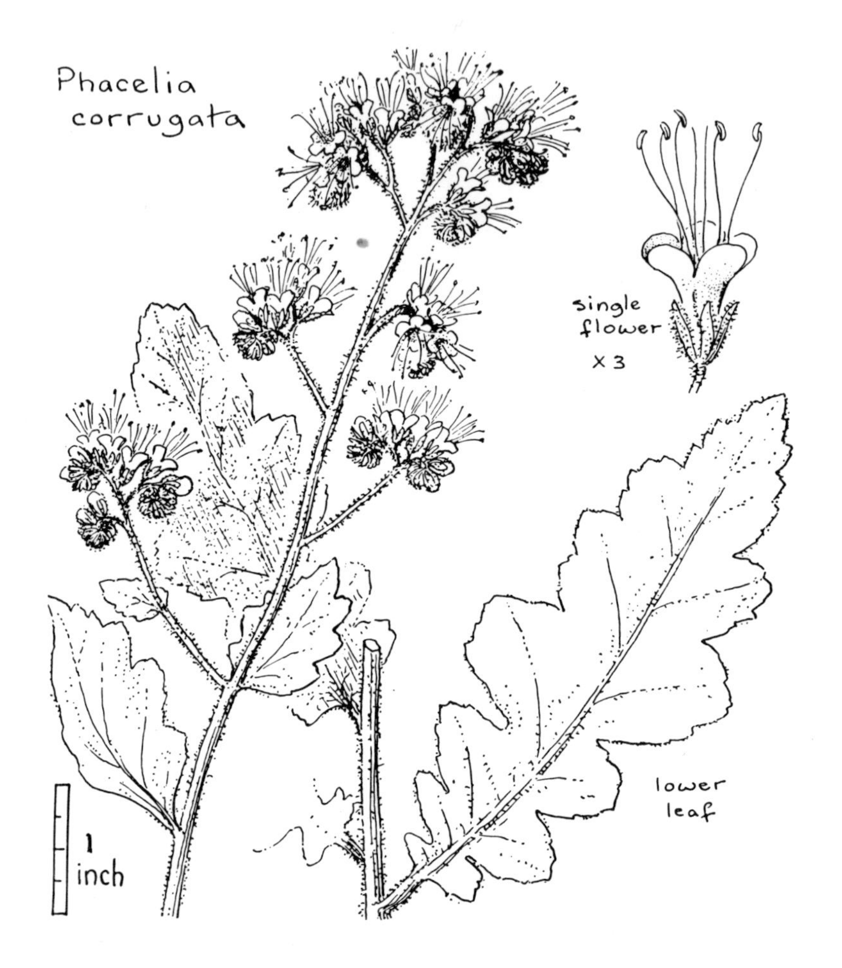

PHACELIA, WILD HELIOTROPE (Phacelia corrugata)

Waterleaf family *(Hydrophyllaceae)*. Blooms lilac-blue, May-September.
Colorado and Utah to Texas, northern Arizona and northern Mexico, 5,000-7,000 feet.

Wild heliotrope is frequently observed on gravelly flats and barren rocky hills of the Pinyon-Juniper Belt. The lilac-blue, bell shaped flowers are striking and are arranged on one side of a curling spike.

The plant is sticky with a disagreeable, onion-like odor, although the flowers themselves have a rather sweet fragrance. The purplish stems have few branches, while leaves have rounded, unequal lobes.

It grows at Grand Canyon along Bright Angel and South Kaibab trails and on the north rim. It also occurs in Aztec Ruins and Bandelier National Monuments.

Another common wild heliotrope *(Phacelia crenulata)* which has a more deeply cleft leaf and darker violet petals, closely resembles the above species, and is found in Zion and Petrified Forest National Parks and in Aztec Ruins, Chaco Canyon, Montezuma Castle and Wupatki National Monuments.

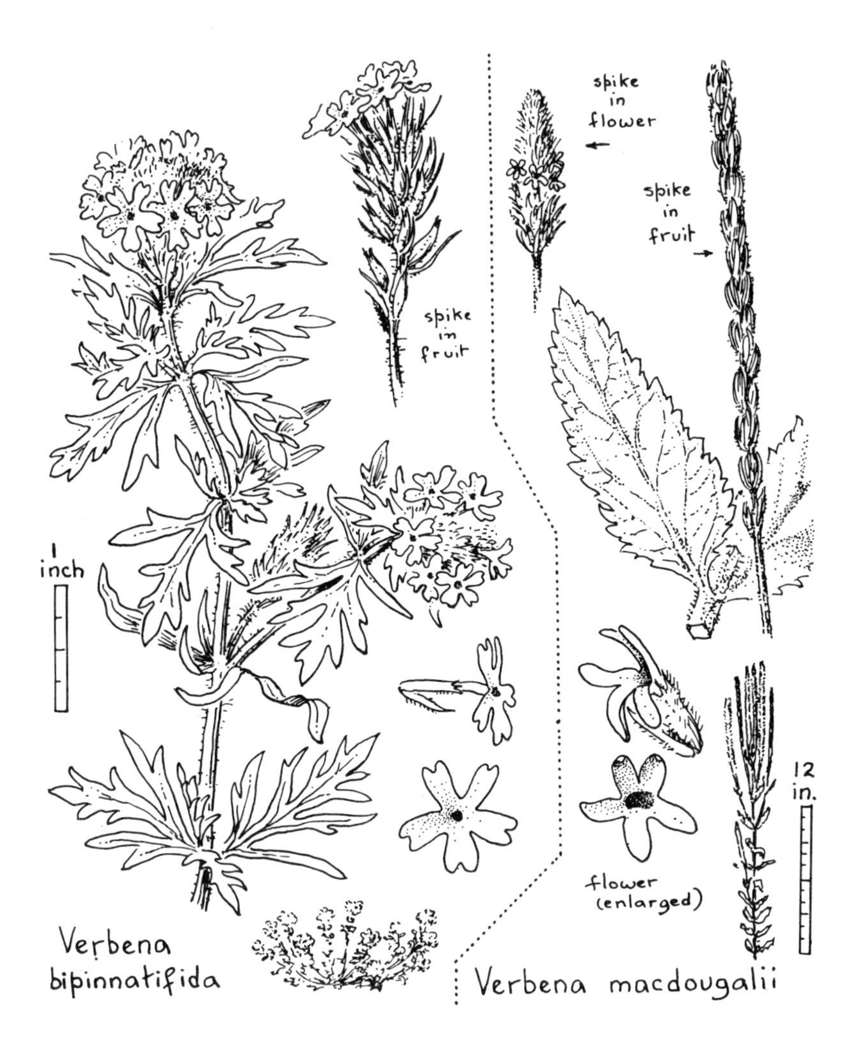

DAKOTA VERBENA, VERBENA, VERVAIN (Verbena bipinnatifida)
Vervain family *(Verbenaceae)*. Blooms lavender purple, May-September.
South Dakota to Alabama, Arizona and northern Mexico, 5,000-10,000 feet.

This is a beautiful and showy plant when it grows in large numbers because of the rich lavender color of the flowers. It is common in open pinyon and ponderosa pine forests. It grows close to the ground, appearing to creep.

SPIKE VERBENA (Verbena macdougalii)
Same family. Blooms violet, June-September.
Southern Wyoming to New Mexico and northern Arizona, 6,500-7,000 feet.

Superficially this plant does not look like a verbena—the square stem, opposite leaves, and two-lipped flowers may mislead one into thinking it is a mint.

The long purple-flowered spikes attain prominence because the plant often grows in large clumps along the road. It is common on the rims of Grand Canyon.

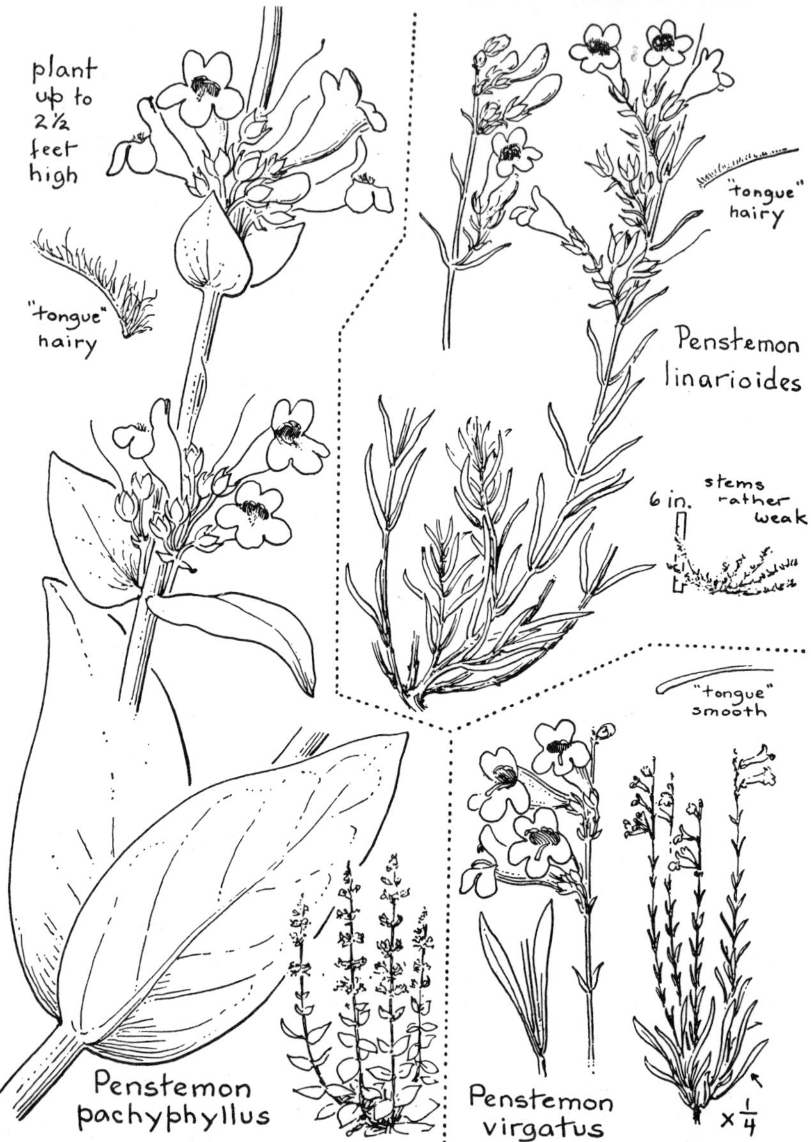

THICKLEAF PENSTEMON, PURPLE PENSTEMON, PURPLE BEARD-TONGUE (Penstemon pachyphyllus)
Snapdragon family *(Scrophulariaceae)*. Blooms deep blue-purple, May-June. Utah, Nevada and northern Arizona, 5,000-7,000 feet.

Coarse pale green leaves.

TOADFLEX PENSTEMON, CREEPING PENSTEMON (Penstemon linarioides)
Blooms blue-purple, June-August.
Colorado, Utah, New Mexico and Arizona, 5,000-9,000 feet.

Thin gray-green leaves. Grows often on limestone. Somewhat prostrate.

WANDBLOOM PENSTEMON, UPRIGHT BLUE PENSTEMON (Penstemon virgatus)
Blooms pale violet with deep purple lines in the throat, summer.
New Mexico and Arizona, 5,000-11,000 feet.

Pink and red penstemons on page 50.

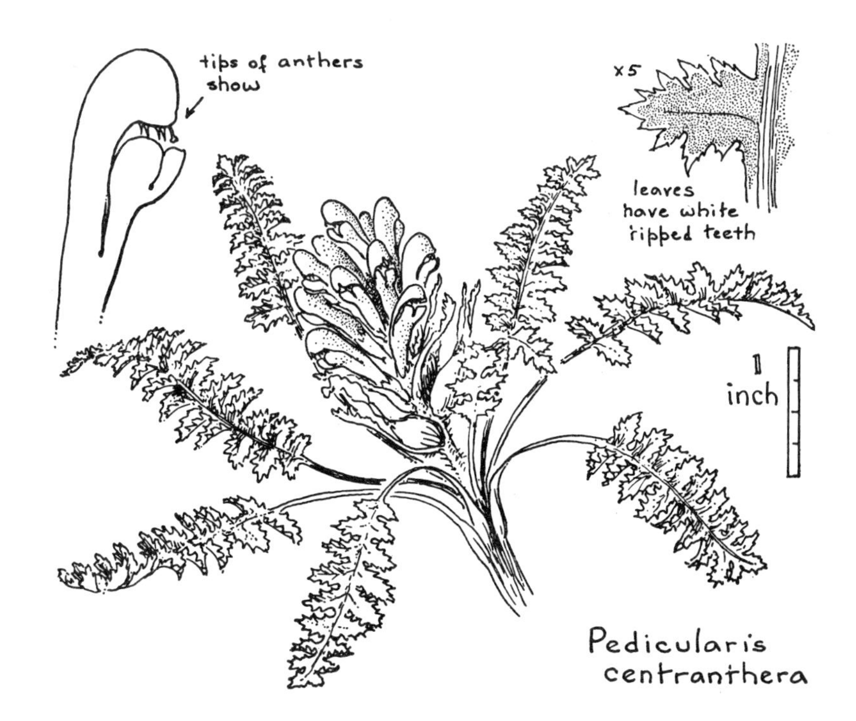

DWARF PEDICULARIS, WOODBETONY, LOUSEWORT
(Pedicularis centranthera)

Snapdragon family *(Scrophulariaceae)*. Blooms white and orchid touched with purple or magenta, April-June.

Colorado, Utah, New Mexico and Arizona, 5,000-7,000 feet.

This particular woodbetony grows under the protection of the pinyon and at slightly higher elevations under the ponderosa pine, blooming in spring and early summer. Partially parasitic on roots of other plants, it is a pretty plant to bear the unpleasant name of "lousewort." Pediculus is a Latin word meaning louse, and the name originated because of the fact that seeds were used, in ancient times, to destroy lice.

The fern-like leaves are finely cut into crinkled lobes and grow in a rosette close to the ground. The teeth of the leaves are tipped with white.

The flowers are pale orchid, nearly white at the base and darker purple or magenta at the top of the lower lip. They crowd close together on a spike that does not exceed the length of the leaves. The flower is two-lipped, the upper lip forming a helmet-shaped hood under which the tips of the stamens appear like small teeth. The lower lip is three-lobed, the center one smaller than the other two.

This species of woodbetony grows at Grand View and elsewhere on both rims at Grand Canyon. In Zion National Park it appears in damp places on Horse Pasture Plateau. At Mesa Verde National Park it can be found under pinyons in the campground. It is also listed at Navajo and Aztec Ruins National Monuments, and in the mountains near Santa Fe.

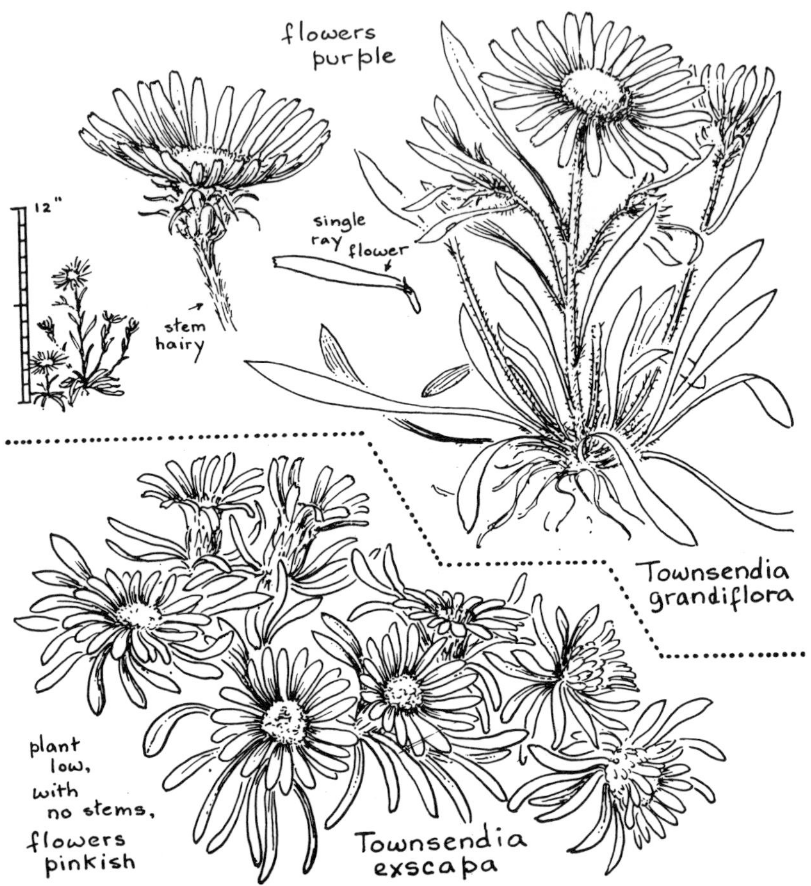

EASTER DAISY, STEMLESS TOWNSENDIA, GROUND DAISY
(Townsendia exscapa)

Sunflower family *(Compositae)*. Blooms white, purplish or pinkish, March-July.

Alberta and Saskatchewan to Texas, Arizona and Chihuahua, 5,000-7,000 feet.

One of the very earliest spring flowers, this low-growing, pale purplish-pink blossom is a welcome sight on dry slopes and mesas, sometimes appearing as early as late February.

It grows in cushion-like tufts and has no stems. The flower nestles in the center of a rosette of gray leaves close to the ground.

The Easter daisy is a common plant with a wide distribution. It grows in the Grand Canyon and Rocky Mountain National Park regions, at Walnut Canyon and Bandelier National Monuments, and is abundant in the hills around Santa Fe.

LARGE-FLOWERED TOWNSENDIA (Townsendia grandiflora)

Blooms violet or purple, June and July.

Wyoming to Nebraska and New Mexico, 5,500-8,000 feet.

This is a similar appearing plant but has larger blossoms and hairy, branching stems that grow about 6 or 8 inches high. It likes open fields and is not so common as the Easter daisy. At Bandelier National Monument it has been seen along the Corral Hill trail.

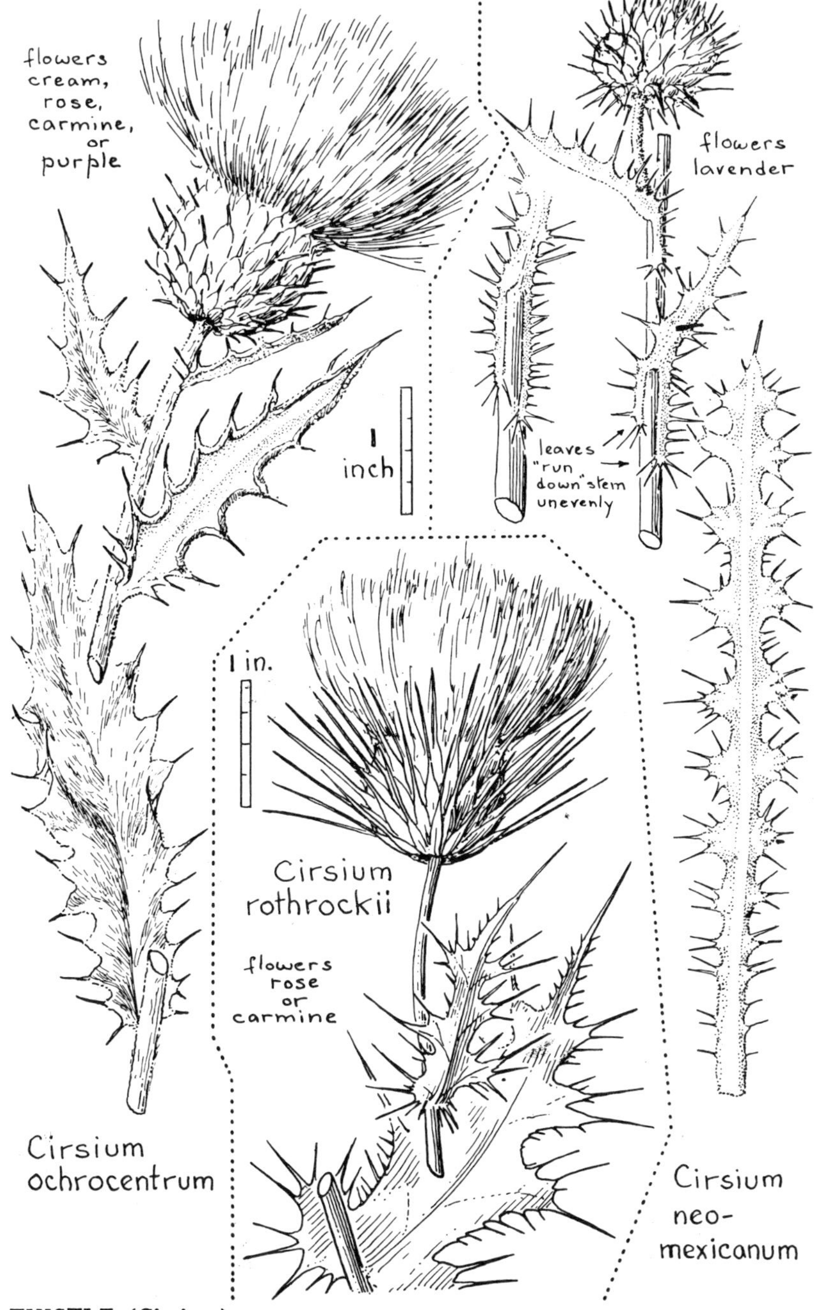

THISTLE (Cirsium)
Sunflower family *(Compositae)*.

NEW MEXICO THISTLE (Cirsium neomexicanum)
Blooms lavender, March-September. Colorado to Nevada, south to New Mexico, Arizona and southern California, 1,200-6,500 feet.

YELLOW-CENTER THISTLE (Cirsium ochrocentrum)
Blooms purplish-rose, often white, May-October, Nebraska to Texas and Arizona, 4,500-8,000 feet.

ROTHROCK THISTLE (Cirsium rothrockii)
Blooms rose-colored to carmine, June-September. Utah and Arizona, 4,000-6,000 feet.

Flowers showy, leaves not woolly.

LITERATURE CONSULTED

ARNBERGER, LESLIE
1947. Flowering Plants and Ferns of Walnut Canyon. *Plateau.* Vol. 20. No. 2. Northern Arizona Society of Science and Art. Museum of Northern Arizona. Flagstaff.

ARMER, LAURA ADAMS
1934. *Cactus.* Frederick A. Stokes Company. New York.

ARMSTRONG, MARGARET
1915. *Field Book of Western Wild Flowers.* G. P. Putnam's Sons. New York and London.

ASHTON, RUTH E.
1933. *Plants of Rocky Mountain National Park.* United States Department of the Interior. National Park Service. United States Government Printing Office. Washington, D. C.

BENSON, LYMAN
1969 *The Cacti of Arizona.* The University of Arizona Press. Tucson. Published by the University of New Mexico Press, Albuquerque.

BREAZEALE, JOHN M.
1930. *Color Schemes of Cacti.* University of Arizona College of Agriculture, Agricultural Experiment Station. University of Arizona Press Tucson, Arizona.

BRITTON, NATHANIEL LORD AND BROWN, HON. ADDISON
1913. *Illustrated Flora of the Northern United States and Canada.* Vol. I, II, and III. Charles Scribner's Sons, New York.

CLEMENTS, FREDERIC E. AND CLEMENTS, EDITH S.
1928. *Rocky Mountain Flowers.* The H. W. Wilson Company, New York.

COULTER, JOHN M. AND NELSON, AVEN
1909. *New Manual of Botany of the Central Rocky Mountains.* American Book Company. New York, Cincinnati, Chicago.

DIXON, HELEN
1935. Ecological Studies on the High Plateaus of Utah. *Botanical Gazette,* Vol. 97, No. 2.

KEARNEY, THOMAS H. AND PEEBLES, ROBERT H.
1960. *Arizona Flora.* University of California Press.

MCDOUGALL, W. M.
1947. *Checklist of Plants of Grand Canyon National Park.* Service Bulletin No. 10. Grand Canyon Natural History Association. Grand Canyon, Arizona.

PESMAN, M. WALTER
1943. *Meet the Natives.* The Smith-Brooks Printing Co. Denver, Colo.

PRESNALL, C. C. AND PATRAW, PAULINE M.
1937. *Plants of Zion National Park.* National Park Service. Zion-Bryce Natural History Association. Zion-Bryce Museum Bulletin No. 1.

PRESTON, RICHARD J.
1947. *Rocky Mountain Trees.* The Iowa State College Press. Ames, Iowa.

RAMALEY, FRANÇIS
1942. *Vegetation of the San Luis Valley in Southern Colorado.* University of Colorado Studies. Series D. Physical and Biological Sciences. Vol. No. 4. Boulder, Colorado.

STOCKWELL, WILLIAM P. AND BREAZEALE, LUCRETIA
1933. *Arizona Cacti.* University of Arizona Bulletin. Biological Science Bulletin No. 1. University of Arizona Press.

TIDESTROM, IVAR
1925. *Flora of Utah and Nevada.* Contributions. United States National Herbarium. Vol. 25.

WOOTON, E. O.
1911. *Cacti of New Mexico.* New Mexico College of Agriculture and Mechanic Arts. Agricultural Experiment Station. Agricultural College, New Mexico. Bulletin No. 78.

WOOTON, E. O. AND STANDLEY, PAUL C.
1915. *Flora of New Mexico.* Contributions. United States National Herbarium. Vol. 19. United States Government Printing Office.

INDEX

Abronia nana 39
Aceraceae 14, 15
Acer grandidentatum 15
Acer negundo 14
Actinea 84
Agave, Utah 56
Agave utahensis 56
Alderleaf mountain-mahogany 65
Allium acuminatum 37
Allium cernuum 37
Allium macropetalum 37
Allium palmeri 37
Amaryllidaceae 56
Amaryllis family 56
Amelanchier 42
Amelanchier utahensis 25
Anacardiaceae 67
Apache-plume 26
Aquilegia chrysantha 58-9
Aquilegia elegantula 40
Aquilegia triternata 40
Arabis pulchra var. pallens 41
Arctostaphylos patula 47-8
Arctostaphylos pungens 46-7
Arctostaphylos uva-ursi 47
Argemone platyceras var. hispida 20
Arizona pincushion cactus 46
Artemisia tridentata 6, 91
Ash 16
Ash-leaved maple 14
Aster arenosus 35
Aster hirtifolius 35
Aster leucelene 35
Astragalus oophorus 28, 95
Astragalus sp. 28
Atriplex canescens 6, 58
Baby aster 35
Babywhite aster 35
Bahia dissecta 83
Banana yucca 18
Barberry family 60
Bearberry 47
Beardtongue 50, 104
Beebalm 101
Beebalm, pony 32
Beech family 13
Beehive cactus 46
Beeplant 59, 95
Bellflower family 54
Berberidaceae 60
Berberis fremontii 60
Berberis repens 60
Bergamot 32, 101
Bigtooth maple 15
Big sagebrush 91
Biting cactus 100
Blackbrush 56
Black sage 91
Bladderpod 63
Bladderstem 57
Blanketflower 85
Blazing-star 68
Bleedingheart 61
Bluebonnet 97
Blue flax 99
Blue gilia 49, 99
Blue lupine 97
Borage family 31-2, 73
Boraginaceae 31-2, 73
Boxelder 14
Branching fleabane 36
Breadroot 98
Brickellbush 34
Brickellia 34
Brickellia grandiflora 34
Brigham-tea 87
Bright paintbrush 52
Broom Snakeweed 74
Buckbrush 27, 29
Buckhorn cholla 71
Buckthorn family 29
Buffaloberry 29, 90
Buffalo currant 64
Buckwheat family 38, 57
Bush morning-glory 7, 101
Buttercup family 19, 40, 58-9, 94
Cactus, beehive 46
Cactaceae 45-6, 69-71, 100
Cactus family 45-6, 69-71, 100
Cacti 7
California redbud 96
Calcchortus nuttallii 17
Calochortus nuttallii var. aureus 17
Campanulaceae 54
Candelabrum cactus 100
Cane cactus 100
Caper family 59, 95
Capparidaceae 59, 95
Caprifoliaceae 53
Cardinalflower 54
Cashew family 67
Castilleja 51-2
Castilleja chromosa 52
Castilleja integra 51-2
Castilleja linariaefolia 52
Ceanothus, desert 29
Ceanothus fendleri 29
Ceanothus greggii 29
Ceanothus martinii 29
Century plant 56
Cercis occidentalis 96
Cercocarpus 6, 65
Cercocarpus ledifolius 65
Cercocarpus montanus 65
Chamaebatiaria millefolium 23
Chamisa 6, 77
Chamiso 58
Chenopodiaceae 58, 89
Cherrystone juniper 10
Chicobush 89
Cholla 70, 100
Chrysothamnus 6
Chrysothamnus nauseosus 77
Chrysothamnus nauseosus var. bigelovii 77
Chrysothamnus nauseosus var. graveolens 77
Cirsium 107
Cirsium neomexicanum 107
Cirsium ochrocentrum 107
Cirsium rothrockii 107
Claret cup, white spined 45
Claytonia lanceolata 40
Claytonia rosea 40
Clematis 19
Clematis ligusticifolia 19
Cleome lutea 59
Cleome serrulata 95
Cliffbush 21
Cliffdwellers candlestick 32
Cliffrose 26-7
Cliff fendler bush 22
Coleogyne ramosissima 56
Colorado juniper 9
Columbine 40, 58-9
Commelinaceae 92
Compositae 34-6, 55, 74-86, 91, 106-7
Coneflower 79
Convolvulaceae 30-1, 101
Convolvulus arvensis 30-1
Convolvulus family 30-1, 101
Corydalis aurea 61
Corydalis montana 61
Coryphantha vivipara var. arizonica 46
Cottonwood 12
Cowania mexicana 27
Cranesbill 43
Creeping penstemon 104
Creeping primrose 30
Crimson monkeyflower 51
Cruciferae 41, 62-3
Cryptantha 73
Cryptantha elata 32
Cryptantha confertiflora 73
Cupressaceae 9-10
Curlleaf Mountain-mahogany 65

INDEX (continued)

Cypress family 9-10
Daisy 36
Dakota verbena 103
Datil yucca 18
Datura meteloides 33
Deerbrowse 65
Delphinium 94
Delphinium menziesi 94
Delphinium scaposum 94
Desert ceanothus 29
Desert pink 55
Desertplume 62
Desertsweet 23
Deserttrumpet 57
Double bladderpod 63
Dwarf pedicularis 105
Easter daisy 106
Echinocereus coccineus 45
Echinocereus triglochidiatus
var melanacanthus 45
Echinocereus viridiflorus 71
Elaeagnaceae 90
Ephedra 7, 87
Ephedra nevadensis 87
Ephedra torreyana 87
Ephedra viridis 87
Ephedraceae 87
Epipactis gigantea 88
Ericaceae 47
Erigeron divergens 36
Erigeron flagellaris 36
Erigeron nudiflorus 36
Eriogonum 38, 57
Eriogonum cognatum 57
Eriogonum inflatum 57
Eriogonum simpsonii 38
Eriogonum stellatum 57
Eriogonum umbellatum 57
Eriogonum wrightii 38
Evening primrose family 30
Fagaceae 13
Fallugia paradoxa 26
Felty groundsel 86
Fendler ceanothus 29
Fendlera rupicola 22
Fendler bush, cliff 22
Fernbush 23
Field bindweed 30-1
Fineleaf yucca 18
Firewheel 85
Flat-topped goldenrod 75
Flax family 66, 99
Fleabane 36
Flowering ash 16
Four-o'clock family 39, 93
Fourwing saltbush 58
Fraxinus anomala 16
Fraxinus cuspidata var. macropetala 16
Fremont barberry 60
Fremont cottonwood 12
Gaillardia 85
Gaillardia pinnatifida 85
Gaillardia pulchella 85
Gambel oak 6, 13
Geraniaceae 43
Geranium caespitosum 43
Geranium family 43
Gilia 49, 99
Gilia aggregata 49
Gila aggregata var. arizonica 49
Gilia longiflora 99
Globemallow 44
Gooseberry globemallow 44
Goosefoot family 58, 89
Golden crownbeard 7, 82
Golden corydalis 61
Golden currant 64
Goldenrod 75
Golden segolily 17
Golden smoke 61
Goldenweed 76
Goldweed 82
Greasewood 6, 89
Green-flowered torch cactus 71
Greenleaf manzanita 48
Groundcherry 72
Ground daisy 106
Groundsel 86
Gutierrezia microcephala 74
Gutierrezia sarothrae 74
Haplopappus gracilis 76
Heart twister 45
Heather family 47
Heath-leaved aster 35
Hedgehog cactus 45
Hedgehog pricklypear 70
Hedgehog, varied 71
Helianthus annuus 81
Helianthus petiolaris 82
Heliotrope 31, 102
Heliotropium convolvulaceum 31
Helleborine 88
Hollygrape 60
Honeysuckle family 53
Horsemint 32, 101
Hydrophyllaceae 102
Hymenoxys argentea 84
Hymenoxys richardsonii
var. floribunda 84
Indian breadroot 98
Indian paintbrush 51
Ipomoea leptophylla 7, 101
Jamesia americana 21
Jointfir family 87
Judas-tree 96
Juniper 6, 9-11
Juniper mistletoe 11
Juniperus monosperma 10
Juniperus osteosperma 10
Juniperus scopulorum 6, 9
Kansas sunflower 81
Kinnikinnic 47
Labiatae 32, 101
Large-flowered townsendia 106
Larkspur 94
Lead bush 90
Leguminosae 28, 66, 95-8
Lesquerella intermedia 63
Lettuce, wire 55
Liliaceae 17, 18, 37
Lily family 17, 18, 37
Linaceae 66, 99
Linum lewisii 99
Linum puberulum 66
Linum usitatissimum 66
Loasaceae 68
Loasa family 68
Lobelia 54
Lobelia cardinalis 54
Locoweed 28, 95
Loranthaceae 11
Lousewort 105
Lupine 97
Lupinus aduncus 97
Lupinus palmeri 97
Mallow family 44
Malvaceae 44
Many-flowered viguiera 80
Manzanita 46-8
Maple family 14, 15
Mariposa lily 17
Matchbrush 74
Mayflower 40
Melilotus albus 28, 66
Melilotus officinalis 66
Mentzelia pumila var. multiflora 68
Mescal 56
Milkvetch 28, 95
Mimulus cardinalis 51
Mimulus guttatus 73
Mint family 32, 101, 103
Mirabilis multiflora 93
Mistletoe family 11
Mockorange 21-2
Monarda menthaefolia 101
Monarda pectinata 32
Monkeyflower 51, 73
Moon lily 33
Mormon-tea 7, 87
Morning-glory 101

INDEX (continued)

Mountain-mahogany 6, 65
Mustard family 41, 62-3
Narrowleaf sunflower 82
Narrow-leaved paintbrush 52
New Jersey tea 29
New Mexico thistle 107
Nodding onion 37
Nut pine 8
Nyctaginaceae 39, 93
Oenothera caespitosa 30
Oenothera runcinata 30
Oenothera runcinata var. leucotricha 30
Oleaceae 16
Oleaster family 90
Olive family 16
Onagraceae 30
One-seeded juniper 10
Opuntia acanthocarpa 71
Opuntia compressa var. macrorhiza 69
Opuntia erinacea var. utahensis 70
Opuntia erinacea var. xanthostemma 70
Opuntia imbricata 100
Orchidaceae 83
Orchis family 88
Oregon grape 60
Painted cup 51-52
Palmer penstemon 50
Papaveraceae 20, 61
Pea family 28, 66, 95-8
Pedicularis centranthera 105
Pelargonium 43
Penstemon 50, 104
Penstemon eatoni 50
Penstemon linarioides 104
Penstemon pachyphyllus 104
Penstemon palmeri 50
Penstemon virgatus 104
Peraphyllum ramosissimum 42
Perky sue 84
Petrophytum caespitosum 24
Phacelia corrugata 102
Phacelia crenulata 102
Philadelphus microphyllus 21-2
Phlox family 48-9, 99
Phlox heliotrope 31
Phlox longifolia 48
Phlox nana 48
Phoradendron juniperinum 11
Physalis fendleri 72
Physaria didymocarpa 63
Pinaceae 8
Pincushion cactus 46
Pine family 8
Pink buckwheat 38
Pink phlox 48
Pink wild snapdragon 50
Pinnate leaved gaillardia 85
Pinus edulis 8
Pinus monophylla 8
Pinyon 6, 8
Pointleaf manzanita 46-7
Poison ivy 67
Polemoniaceae 48-9, 99
Polygonaceae 38, 57
Pony beebalm 32
Poppy family 20, 61
Populus fremontii 12
Portulacaceae 40
Portulaca family 40
Potato family 33, 72
Prairie flax 99
Prairie spiderwort 92
Prairie sunflower 82
Prickle-poppy 20
Pricklypear 69, 70
Princes plume 62
Psoralea lanceolata 98
Psoralea tenuiflora 98
Purple beardtongue 104
Purple penstemon 104
Quercus gambelii 6, 13
Quercus turbinella 13
Quinine bush 27
Rabbitbrush 6, 77
Ragleaf bahia 83
Ragwort 86
Ranunculaceae 19, 40, 58-9, 94
Ratibida columnaris 79
Rattleweed 95
Redbud 96
Red columbine 40
Red-flowered pricklypear 70
Red gilia 49
Red penstemon 50
Rhamnaceae 29
Rhododendron 47
Rhus radicans 67
Rhus trilobata 67
Ribes aureum 64
Ribes leptanthum 64
Rockcress 41
Rockmat 24
Rockrose 24
Rock goldenrod 75
Rock spiraea 24
Rocky mountain beeplant 95
Rocky mountain juniper 6, 9
Rocky mountain white oak 13
Rocky mountain zinnia 78
Rosaceae 23-7, 42, 65
Rose family 23-7, 42, 65
Rothrock thistle 107
Roundleaf buffaloberry 90
Rubber plant 84
Rubber rabbitbrush 77
Sacred datura 33
Sagebrush 6-7, 91
Salicaceae 12
Saltbush 58
Sand-verbena 39
Sarcobatus vermiculatus 6, 89
Saxifragaceae 21-2, 64
Saxifrage family 21-2, 64
Scarlet bugler 50
Scrophulariaceae 50-1, 73, 104-5
Scrub maple 15
Scurf-pea 98
Segolily mariposa 17
Senecio longilobus 86
Senecio multicapitatus 86
Serviceberry 25, 42
Shadbush 25, 58
Shad-scale 6, 58
Sheath flower 34
Shepherdia rotundifolia 90
Shrub live oak 13
Singleleaf ash 16
Singleleaf pinyon 8
Skeleton weed 55
Skunkbush 67
Skunkweed 95
Skyrocket gilia 49
Small-leaved globemallow 44
Small snakeweed 74
Snakeweed, broom 74
Snakeweed, threadleaf 74
Snapdragon family 40, 50-1, 73, 104-5
Snowberry, mountain 53
Soapweed 18
Solanaceae 33, 72
Solidago petradoria 75
Solidago sparsiflora 75
Sore-eye-poppy 44
Spanish-bayonet 18
Sphaeralcea 44
Sphaeralcea grossulariaefolia var. pedata 44
Sphaeralcea parvifolia 44
Spiderflower, yellow 59
Spiderwort family 92
Spike verbena 103
Sprawling daisy 36
Spreading daisy 36
Spring-beauty 40
Squawapple 42
Squawbush 67
Stemless townsendia 106
Stanleya pinnata 62
Stemless primrose 30
Stephanomeria pauciflora 55

INDEX (continued)

Stephanomeria tenuifolia 55
Stickleaf 68
Strawberry tomato 72
Stream orchis 88
Sulphurflower 57
Sulphur eriogonum 57
Sumac family 67
Sunflower family 34-6, 55, 74-86, 91, 106-7
Symphoricarpos oreophilus 53
Syringa 21
Tansybush 23, 25
Tassel-flowered brickellia 34
Thickleaf penstemon 104
Thistle 107
Thistle poppy 20
Thornapple 33
Threadleaf groundsel 86
Threadleaf snakeweed 74
Toadflex penstemon 104
Townsendia exscapa 106
Townsendia grandiflora 106
Tradescantia occidentalis 92
Tradescantia pinetorum 92
Tradescantia scopulorum 92
Trumpet gooseberry 64
Tuberous pricklypear 69
Twinpod 63
Upright blue penstemon 104
Utah agave 56
Utah juniper 10
Varied hedgehog 71
Verbena 39, 103
Verbena bipinnatifida 103
Verbenaceae 103
Verbena macdougalii 103
Verbesina enceliodes 7, 82
Vervain family 39, 103
Viguiera 80
Viguiera multiflora 80
Viguiera multiflora var. nevadensis 80
Wandleaf penstemon 104
Waterleaf family 102
Waxflower 21
Weeping juniper 9
Western redbud 96
Western virginsbower 19
White horsemint 32, 101
White scurfpea 98
White spined claret cup 45
White sweetclover 28, 66
Wholeleaf painted cup 51-2
Wild buckwheat 38, 57
Wild chrysanthemum 83
Wild crab apple 42
Wild geranium 43
Wild heliotrope 102
Wild lilac 29
Wild onion 37
Wild snapdragon 50, 104
Wild zinnia 78
Willow family 12
Wire lettuce 55
Wolfberry 53
Woodbetony 105
Wright buckwheat 38
Wyoming painted cup 52
Yellow beeplant 59
Yellow borage 73
Yellow columbine 40, 58-9
Yellow coneflower 79
Yellow daisy 76
Yellow flax 66
Yellow-center thistle 107
Yellow-flowered cane cactus 71
Yellow-flowered evening-primrose 30
Yellow-flowered pricklypear 69
Yellow monkeyflower 51, 73
Yellow spiderflower 59
Yellow sweetclover 28, 66
Yucca 7, 18, 56
Yucca baccata 18
Yucca, banana 18
Yucca angustissima 18
Yucca, datil 18
Yucca, fineleaf 18
Zinnia 78
Zinnia elegans 78
Zinnia grandiflora 78
Zinnia pumila 78